Progressive Energy Policy

Series Editors

Progressive Energy Policy is a new series that seeks to be pivotal in nature and improve our understanding of the role of energy policy within processes of sustainable, secure and equitable energy transformations. The series brings together authors from a variety of academic disciplines, as well as geographic locations, to reveal in greater detail the complexities and possibilities of governing for change in energy systems. Each title in this series will communicate to academic as well as policymaking audiences key research findings designed to develop understandings of energy transformations but also about the role of policy in facilitating and supporting innovative change. Individual titles will often be theoretically informed but will always be firmly evidence-based seeking to link theory and policy to outcomes and changing practices. Progressive Energy Policy is focussed on whole energy systems not stand alone issues; inter-connections within and between systems; and on analyses that moves beyond description to evaluate and unpack energy governance systems and decisions.

More information about this series at
https://link.springer.com/bookseries/15052

Catherine Butler

Energy Poverty, Practice, and Policy

palgrave
macmillan

Catherine Butler
University of Exeter
Exeter, UK

Progressive Energy Policy
ISBN 978-3-030-99431-0 ISBN 978-3-030-99432-7 (eBook)
https://doi.org/10.1007/978-3-030-99432-7

This Palgrave Macmillan imprint is published by the registered company Springer Nature Switzerland AG
The registered company address is: Gewerbestrasse 11, 6330 Cham, Switzerland

Acknowledgements

First and foremost, thanks go to my husband Ian Wilson for his unending support in all things, without which this book would never have been completed. I would like to thank Dr. Karen Parkhill for working with me (amid much upheaval and change) in developing and delivering the research project on which this book is based and for her hugely important, thoughtful, and constructive comments on an earlier draft. Similarly, thanks go to Sarah Royston for incredibly insightful, helpful, and kind comments on an even earlier version. Big thanks go to Karen Bickerstaff, Gordon Walker, and Paulina Luzecka for their contributions to the research and project, and for generally being inspiring, excellent people to work with. Sincere thanks also go to the research participants who gave up their time and insights as part of the project. And thanks go to Rachael Ballard, Aishwarya Balachandar, and Helen van der Stelt at Palgrave-Macmillan for sticking with me for far longer than I (and I'm sure they) had ever imagined! Finally, the research was supported by UK Research and Innovation (UKRI) under the Engineering and Physical Sciences Research Council award EP/M008150/1.

About This Book

Energy poverty represents a pressing policy issue that intersects with wider concerns about inequality in society and cuts across different areas of governance. This book examines the implications of welfare policy for energy poverty, engaging with key academic conceptual debates at the forefront of energy demand research. These debates concern academic research that has given focus to the multiple factors that create and shape experiences of energy poverty. And work that has examined processes through which everyday social practices require increasing levels of energy use with impacts on sustainability. This book develops an analysis that reveals how novel insights can be made visible through combining these different ways of thinking about energy demand issues. It presents a distinctive approach to examining energy poverty that places inequalities at the heart of debates about the advancing energy intensity of contemporary societies. In doing so, it contributes to the frontiers of energy poverty research and responds to critiques of social-practice-informed analyses of energy demand that highlight the limited attention given to inequalities within such work.

Contents

About the Author

Dr. Catherine Butler is Senior Lecturer in Human Geography at University of Exeter. Her research centres around analysis of environmental governance processes with focus on the intersections between policy, politics, and everyday life. She has published extensively on topics including energy transitions in everyday life, behavioural change and social practice, wellbeing impacts of floods, and governance of climate adaptation. This book arises out of her four-year EPSRC (Engineering and Physical Sciences Research Council) funded project that looked at how welfare policy has had implications for issues of energy demand.

Abbreviations

BEIS	Department for Business, Energy, and Industrial Strategy
CERT	Carbon Emissions Reduction Target
CESP	Community Energy Saving Programme
COP	Conference of the Parties
CRC	Carbon Reduction Commitment
DECC	Department of Energy and Climate Change
DfT	Department for Transport
DLA	Disability Living Allowance
DWP	Department for Work and Pensions
ECO	Energy Company Obligation
EPC	Energy Performance Certificate
ESA	Employment and Support Allowance
HMO	House in Multiple Occupation
IHD	In-Home Display
JSA	Jobseeker's Allowance
LIHC	Low Income High Costs
LILEE	Low Income Low Energy Efficiency
OFGEM	Office of Gas and Electricity Markets
PIP	Personal Independence Payment
PRS	Private Rented Sector
SMIP	Smart Metering Implementation Programme
ZCH	Zero Carbon Homes

CHAPTER 1

Introduction

Abstract This introductory chapter outlines the book's central concern with energy poverty and sets the context for the arguments addressed and advanced through its pages. A significant existing literature has examined issues of energy poverty, with key interventions using concepts of energy vulnerability, precarity, and capabilities. But an equally large body of work has examined problems of reducing energy demand by focusing on the importance of reshaping and shifting practices. Though these two major literatures on energy demand issues occasionally intersect, they have rarely been brought into direct conversation with one another. The book examines issues of energy poverty with focus on advancing conceptual debates by engaging with ideas that span these two areas, principally those concerning capabilities, precarity, and practice. In doing so, it contributes to the frontiers of energy poverty research and responds to critiques of practice-informed analyses of energy demand that highlight the limited attention given to inequalities within such work.

Keywords Energy poverty · Capabilities · Practice theory · Invisible energy policy · Energy needs

C. Butler, *Energy Poverty, Practice, and Policy*, Progressive Energy Policy,
https://doi.org/10.1007/978-3-030-99432-7_1

Energy Poverty, Practice, and (Invisible Energy) Policy

Energy poverty refers to issues that span access to energy, energy deprivation, and its under-use within daily life. Broadly, the central concern of research and policy in this space is with the negative outcomes that a lack of energy use has for wellbeing. Within the UK and many other global contexts, the focus has often been on those that cannot afford to heat their homes, with solutions posed as ones principally involving building efficiency and to a lesser extent affordability. Within the academic literature, however, there has been increasing recognition of the complex and multifaceted set of issues involved in both creating and addressing problems of energy poverty. This body of work has made key conceptual interventions that put forward important arguments about how energy poverty can, or even should be, understood. These concern centrally a shift beyond a focus on heat, efficiency, and affordability to take in a much wider range of issues, dimensions, and dynamics that are important in shaping energy poverty.

One such area of conceptual development has been in the move from understanding energy poverty as a static state that a person is either 'in' or not, to analysis of the conditions that shape or lead to such experiences—termed energy vulnerability (see Bouzarovski, 2018). Though this step within understanding has been extremely important, it has also been critiqued for displacing focus from the wider structural and social processes implicated in creating conditions of energy poverty. This is because vulnerabilities research tends to examine the characteristics and capacities of the person (such as whether someone is disabled, young, elderly, low income, and so forth), consequently individualising the causes of energy poverty (Middlemiss & Gillard, 2015).

Moving beyond this, then, energy poverty scholars, such as Bouzarovski and Petrova (2015), Petrova (2018), Day et al. (2016), Simcock et al. (2016), and Middlemiss et al. (2019), have worked with concepts of energy services, precarity, and capabilities offering a stronger basis for analysis of the socio-political dynamics shaping experiences of energy poverty. This work marks several important shifts within understandings of energy poverty. First, Bouzarovski and Petrova (2015) have engaged the concept of energy services to bring focus on the benefits that people derive from using energy, such as mobility, lighting, cooking, and so on, rather than the energy itself. This entails recognition of the ways that domestic energy poverty is only fully understood by looking across multiple energy services and their interconnections, as well as taking analysis beyond the confines of the home. It also brings focus on the ways that fulfilment of energy needs underpins many of the 'functionings' that enable people to have a (minimally) decent quality of life (Bouzarovski & Petrova, 2015, p. 34; see also Simcock et al., 2016).

This work highlights the importance of examining the driving forces of energy poverty in terms of the ways that abilities to meet energy service needs are affected by multiple factors. Such factors include vulnerabilities but also encompass the 'concatenation of activities, infrastructures, and resources necessary to provide households with energy' (Bouzarovski & Petrova, 2015, p. 35). A concern with provisioning thus brings into view the combination of social, economic, political, and infrastructural factors that contribute to people being in positions of energy poverty.

Second, a set of interventions have built on this to bring conceptual focus on the capabilities framework as a way to create a more sophisticated understanding of energy poverty (e.g. Day et al., 2016; Middlemiss et al., 2019). This draws centrally on the work of Sen (1999) and Nussbaum (2011) where 'capabilities' to support opportunities for functioning and achieving wellbeing are argued to be central to human development. Capabilities include things like the ability to secure healthy food, or to engage politically, or secure income. This framework, and its attendant concepts, has been advanced within the energy poverty literature as a key way to engage with more complex understandings of the issues. Central to this is recognition that many capabilities are underpinned by or related to various energy services. Starting from the capabilities that energy affords offers a conceptual approach that can draw in the wide range of human needs that energy is utilised in meeting and, therefore, the multiple social and political processes involved in the conditioning of energy poverty. Energy poverty analysis has, then, come a long way from a narrow focus on issues of heat, efficiency, and cost, moving towards approaches that emphasise *what energy is for* and recognise the complexities of the dynamics shaping both its causes and the nature of lived experiences.

Within these conceptual developments, which have come to shape debates about energy poverty, scholars make occasional references across to the major theoretical advances arising from a wider energy demand research agenda that is grounded in practice theory (e.g. Bouzarovski &

Petrova, 2015; Day et al., 2016). The practice theory-inspired literature has been primarily concerned with issues of environmental sustainability related to energy demand, mounting an important critique of behavioural or efficiency approaches within policy (e.g. Shove, 2010, 2017). This critique asserts that such approaches are highly limited because they fail to engage with more fundamental questions concerning how our particular requirements for energy are constructed and reproduced (e.g. see Shove, 2003; Shove et al., 2012).

Shove (2003), amongst others, has asserted that rather than focusing on improving the efficiency of technologies that support and engender particular kinds of practices, energies should be directed towards examining the specification of need and the processes by which various forms of demand come to be considered normal. Equally, to understand why people do or do not buy more efficient technologies or drive more frequently, the challenge becomes one of 'understanding the collective transformation of convention and hence the dynamics of energy demand' (Shove, 2004: 1055). This approach to thinking about energy demand gives cause to examine the ways that everyday practice is shaped and comes to be seen as normal. In this, analyses have demonstrated the role of government objectives, investments, and ways of working in shaping social practice and, in doing so, constituting the need for energy in the home, at work, and in moving around.

Indeed, an emerging area of practice theory-inspired energy research has sought to explicitly examine the role of government strategies, policies, and processes in shaping practices with implications both for how we understand problems of energy demand and for how we define the parameters of relevant governance (Butler et al., 2018; Cox et al., 2019; Gormally et al., 2019; Greene & Fahy, 2020; Royston et al., 2018). A central claim of this work is that requirements for energy are shaped, shifted, and constituted through a wide range of different intersecting policy areas, reaching far beyond energy policy *per se*. This body of emerging work has coined the terms 'invisible' or 'non-energy' policy as ways to characterise the analytic endeavour. A range of policy areas have been addressed under this remit, including education (Gormally et al., 2019; Royston, 2016), health (Blue, 2017), digital communications (Morley et al., 2018), and in my own work, welfare policy (Butler et al., 2018). These analyses have worked to show how different policy areas have implications for energy demand and related issues.

These key ideas from within practice theory-based energy research have been developed within multiple studies addressed at the environmental sustainability implications of energy demand, but rarely have they been used to think about energy poverty. Conversely, a key critique of practice theory-based analyses of energy demand cites the inadequacy of such approaches to account for inequality and power relations (Walker, 2013). Though not inherent to wider practice theory (cf. Bourdieu, 1998; Foucault, 1991), relations of power and inequality have been largely neglected within practice-inspired energy demand research. A few existing conceptual contributions advance some areas of overlap, exploring inequality in discussions of practice theory and energy (e.g. Shove, 2002; Shove et al., 2012; Walker, 2013). But beyond this, these issues still remain largely unaddressed.

Within the energy poverty literature, the existing engagement with practice theory or practices (e.g. Bouzarovski & Petrova, 2015; Day et al., 2016; Petrova, 2018) has signalled some clear touchstones for the ways that practice-theory thinking can be relevant for understanding energy poverty. For example, it has been influential in discussions about the importance of focusing on the practices for which energy is used, rather than energy itself. But this has not entailed a more detailed analytic endeavour to look across practice-theory conceptual work or engage in a deeper conversation with the fields of inquiry it has inspired. Both literatures offer central ideas that I argue here could be taken much further to advance understanding of energy poverty, as well as open-up avenues for future practice-based energy research that can better attune to relations of inequality.

This book is ambitious in seeking to explore how these different distinctive areas of analysis can be further advanced through dialogue with one another, developing the conversation along key conceptual lines. These concern, first, an interest in the ways that the invisible energy policy agenda—as a nascent area of practice-inspired energy research—has potential for thinking about energy poverty. In particular, I argue it can have important application for advancing existing energy poverty work that aims to bring the social and political processes shaping energy deprivation into view (cf. Petrova, 2018; Middlemiss, 2016).

For example, research has looked at the ways that energy deprivation is institutionalised and normalised through policy and governance with important implications for political mobilisation (Petrova, 2018). Analysis has also been addressed at the ways that *subjects* of fuel poverty policy are constituted in particular ways compared with other policy areas. For instance, Middlemiss (2016) highlights how fuel poverty subjects tend to be cast as vulnerable and worthy of support, contrasting this with subjects of welfare policy that are often situated as undeserving and as harbouring individual deficits. I argue here that these ideas can be brought into dialogue with those from invisible energy policy to advance the research agenda with power and inequal relations more firmly in view. Though the invisible energy policy agenda offers potential for greater consideration of power, such issues and related conditions of inequality have yet to form a focus. I assert, therefore, that bringing analytic attention within invisible energy policy work onto energy poverty offers distinctive possibilities for developing future analysis across both areas.

Second, I bring a focus on the constitution of need and social reproduction of practice arguing this offers novel routes to further understand how energy deprivation is created and how it might be addressed. Working with these conceptual ideas takes analysis beyond looking at how abilities to *meet needs* are affected by wider social and political dynamics, to consider the processes through which those needs are actively constituted in the first place. Importantly, this brings into view the implications of the advancing energy intensity of societies for energy deprivation. Further, I show how by exploring the constitution of need in relation to energy poverty, insights can be opened up into how relations of inequality and power shape processes of social reproduction with implications for practice theory-based energy research. Centrally, these insights concern the specific ways that power relations operate through governance to differentially shape people's agency to resist, negotiate, and enact practices.

The book thus draws a line from practice theory-inspired ideas, arising from work on transitions and invisible energy policy, through to energy poverty analysis, wherein debates about vulnerability, precarity, and capabilities are advanced. In addition to generating insights relevant to both these areas of conceptual and empirical analysis, the book also intersects with the wider theoretical traditions on which these literatures draw, centrally practice theory, wellbeing, and capabilities. Though these different conceptual areas are well worn within energy demand and energy poverty research respectively, they are rarely integrated. For example, wellbeing is already important in debates about energy poverty but has been less central to debates about the constitution of need that are core to practice theory-based analyses of energy. In threading these different conceptual areas together and working through them with an empirical analysis, the book seeks to make a contribution to theory that has a wider relevance for those outside of energy research.

The Empirical Study: Methods and Approach

The contributions of this book are developed through engagement with an empirical study that looked at UK welfare policy as an area of invisible or non-energy policy (see also Butler et al., 2018). Welfare policy represents a core part of governance systems for many countries and territories. It has a long history as a part of governance arrangements in capitalist democracies and has often been contentious forming a focus for ideologically driven political battles. Centrally, these battles concern how societies should or should not tackle inequality and are underpinned by beliefs about what creates inequality in the first place. In the UK, in particular, welfare and employment policy represents one of the government's highest expenditure areas and is frequently the focus of public debate and media attention. The importance of this policy area combined with its distance from specific energy directives denotes this as an interesting empirical case for examining 'invisible energy policy' or 'non-energy policy' (Royston et al., 2018).

By focusing on an area of policy outside of energy, the locus of analysis is shifted and, I argue, a different way of looking at energy demand issues is revealed. Crucially, it takes analysis beyond definitions and categorisations of energy problems as they are currently formulated within existing governance structures. Here, I use this different orientation and starting point to provide distinctive insights into energy poverty and advance new lines of questioning concerning the processes by which energy demand is re/produced and created.

The three-year (2015–2018) project consisted of methods including document analysis, in-depth qualitative and biographical interviews, and workshops. Each method and the approach adopted within the project are explained here. The project involved detailed analysis of key documentary materials relevant to the project aims. This focused on documents related to contemporary welfare reforms and energy demand policy arising from the two UK government departments with responsibilities in this area—namely the Department for Work and Pensions (DWP) and the Department for Business, Energy and Industrial Strategy (BEIS—and the former Department of Energy and Climate Change [DECC]). Documents analysed included parliamentary speeches, political party election manifestos, reports, strategy documents, presentations, academic literature, and government department websites. The documents selected included ones predating the start of the research project in 2015, spanning back to the early 1980s. However, the focus for the document analysis was primarily on the time-period since 2010 when a Conservative led Government was elected in the UK. This time-period was selected as a focus because it marked the beginnings of major contemporary welfare reforms.

The interviews involved a total of 62 participants with people implicated in different ways in governance across energy and welfare policy. This spanned people involved in; (1) national policy, including civil servants, agencies, and NGOs; (2) those in professional governance roles at city scales, such as within local government, agencies, and charities; and (3) people directly affected by welfare policy within their everyday lives. The research thus takes in different actors implicated in processes of governance, looking far beyond the state. In-depth interviews lasting 1–3 hours were conducted between September 2015 and July 2017 in two phases. A first phase engaged with those in national roles, while a second phase moved to focus on two city case study areas interviewing people working in agencies and organisations implicated in welfare and/or energy policy, as well as those directly affected.

The city case study areas selected were, York in the North-East of England and Bristol in the South-West of England. The interviews with people across these two case study cities offered insight into the ways that national policies are made manifest and negotiated as part of professional roles and everyday lives. While more conventional in-depth qualitative interviewing techniques were employed for those in professional and stakeholder roles, biographical interview techniques were utilised for those directly affected by welfare policies. This approach to interviewing focuses discussion on people's lives and their experiences eliciting narratives that reach backward and forward in time. It can be useful for studies, such as this one, that are seeking to build insight into complexity and often non-linear processes of intersection; in this case between policies and everyday life experiences (see Butler et al., 2014). The relevance of experience-centred approaches to examining governance has been foreshadowed within the literature on non-energy policy and energy demand more generally (e.g. Butler et al., 2014, 2016; Greene & Fahy, 2020).

Such an approach offers insight into the ways that people integrate, respond to, and negotiate policy as part of their everyday lives, as well as bringing social differentiation and power relations more sharply into focus. By looking across these varied experiences and addressing different lines of questioning the project was able to build insight into the intersections between both the personal and political or the public and private, and different areas of policy that, while distinct at national scales, are intimately interwoven within the fabric of people's everyday lives. A final phase of the research involved three workshops (participant $n = 28$) with those working in roles related to welfare policy and/or energy to bring focus on possibilities for change that might arise out of considering invisible energy policy. These were held in 2018 across London, York, and Bristol engaging both national stakeholders and those from our city case sites. They offered further insight into the nature of policy intersections across welfare reform and energy poverty, as well as the potential openings, and constraints, for change.

The book does not develop a complete or exhaustive analysis of the data derived from the project. Rather, the empirical material is used in a more circumscribed way to draw together and engage with the core themes and ideas with which the book is concerned. This entails focus on the intersections of energy poverty and practice-based research, and on interweaving conceptual directions associated with capabilities, the constitution and specification of need, and invisible energy policy. The empirical analysis takes a first step towards realising insights that can be made visible by thinking across these different theoretical developments, and advances possibilities for future inquiry at the intersections. In this, greater focus is given to the biographical interviews with those directly affected by welfare policy in their everyday lives. However, the wider data and study inform and foreground the analysis, for example, by offering understanding of the core policy changes that are reflected in people's accounts of their life experiences.

For ethical purposes, the presentation of quotes as part of the empirical analysis does not attribute them using names of interviewees and/or their organisations. Instead, generic identifiers are used, with interview extracts labelled using the tags of 'biographical interviewee' and 'stakeholder interviewee' (either local or national), as well as the case site location and a participant number. These identifiers are not significant to the analysis *per se* but are used to distinguish between different interviewees.

The Book's Structure

Throughout the book, I develop several areas of contribution that emerge from both conceptual analysis and engagement with the empirical materials outlined above. These contributions are advanced through the book's chapters as follows.

The opening two chapters explore the conceptual lines from within energy demand research that are of central concern in the book, looking at *Poverty* (Chapter 2) and *Practice* (Chapter 3) literatures. Chapter 2 synthesises key conceptual debates at the cutting edge of energy poverty research and contributes to thinking about how a capabilities-based analysis can be taken forward. Chapter 3 discusses key ideas arising from practice theory-inspired work—including those relating to the constitution of demand and invisible energy policy. It reflects on the ways that inequality can be (and has been) thought about within practice-theory energy research, as well as on how bringing energy poverty concerns into focus raises distinctive possibilities and avenues for analysis.

Chapter 4 moves to focus on *Policy.* It reviews the policy landscape relating to energy demand, with focus on the ways that fuel poverty has been defined and addressed. It examines change and continuity in policy over time reviewing past and present initiatives and strategies, and reflects on the gulf that exists between contemporary policy definitions of fuel poverty and academic analysis in this space. It then introduces the welfare policy case, as an area of invisible energy policy, outlining key policy changes and developments within this area that are relevant to the empirical analysis, as well as discussing points of connection and disconnect across to energy policy within this sphere.

The next two chapters (5 & 6) use key examples from the empirical data to show how combining ideas across the energy poverty and practice-based energy demand literatures can be important in bringing to light insights and avenues for further research. Chapter 5 focuses on how the invisible energy policy agenda can be advanced in relation to issues of energy poverty by looking at the case of UK welfare policy and its role in shaping energy deprivation. Building from the capabilities-based understanding of energy poverty, discussed in Chapter 2, the analytic endeavour takes forward thinking on invisible energy policy by going beyond examination of the ways welfare policies more directly affect energy deprivation. Although such direct forms of policy impact are discussed, the analysis moves to look at how wider political discourses that pervade different policy areas (cf. Middlemiss, 2016) are shaping experiences and manifestations of energy poverty in important ways. In particular, it offers insights regarding the normalisation and institutionalisation of energy precarity (see Petrova, 2018), looking at how such processes are intertwined with (non-energy) policy discourses and approaches.

Chapter 6 once again builds from a capabilities-based understanding of energy poverty, but the focus in this chapter is on ideas from practice theory that concern the constitution of need through (invisible energy) policy. It reflects on how arguments concerning the ways that needs come to be specified have relevance for energy poverty research by looking at how needs are created and imposed through welfare policy reform. Crucially, in this chapter focus is brought onto questions not only of whether people can meet needs, but also of how those needs are created, as well as how abilities to resist, negotiate, and constitute needs are also unequal. With emphasis on invisible energy policy, the analysis in the chapter develops deeper understanding of how power relations and inequality figure in the constitution of practices. It does so by making explicit the role of policy in processes of constitution and by highlighting how different policy areas act on their subjects in ways that reflect patterns of inequality. This suggests the importance, then, of a deeper analysis of inequality for practice-based understandings of social action. Finally, Chapter 7 concludes the book with reflection on the implications for energy poverty research, for practice theory and invisible energy policy, and for wider contexts of policy and practice.

And finally... a note on key terms...

The terms fuel poverty and energy poverty are often used to denote problems of energy deprivation across Global North and Global South countries respectively (Bouzarovski & Petrova, 2015). Fuel poverty, as such, often refers more to issues of energy affordability—particularly that associated with heating and cooling—while energy poverty tends to be used in reference to issues of energy access associated with lack of infrastructure and technology. In this book, I use the term fuel poverty to refer to narrower definitions of energy deprivation (e.g. as related primarily to heat), while energy poverty is utilised to refer to wider understandings taking in multiple and diverse uses of energy, including mobility and domestic heating and non-heat uses (e.g. lighting, computing). I also use the terms energy vulnerability, energy precarity, and energy capabilities but these are discussed in detail in Chapter 2. Other key terms include *practice-based* or *inspired energy research* (also see Walker, 2013), which I use to refer to the body of energy research that arises primarily from Shove's (e.g. 2003; with Pantzar and Watson, 2012) conceptual development of practice theory through application to environmental sustainability.

References

Blue, S. (2017). Reducing demand for energy in hospitals: Opportunities for and limits to temporal coordination. In A. Hui, R. Day, & G. Walker (Eds.), *Demanding energy: Space time and change* (pp. 313–338). Palgrave.

Bourdieu, P. (1998). *Practical reason: On the theory of action*. Polity Press.

Bouzarovski, S. (2018). *Energy poverty: (Dis)assembling Europe's infrastructural divide*. Palgrave.

Bouzarovski, S., & Petrova, S. (2015). A global perspective on domestic energy deprivation: Overcoming the energy poverty–fuel poverty binary. *Energy Research & Social Science, 10*, 31–40.

Butler, C., Parkhill, K. A., & Luzecka, P. (2018). Rethinking energy demand governance: Exploring impact beyond "energy" policy. *Energy Research & Social Science, 36*, 70–78.

Butler, C., Parkhill, K. A., & Pidgeon, N. F. (2016). Energy consumption and everyday life: Choice, values and agency through a practice theoretical lens. *Journal of Consumer Culture, 16*(3), 887–907.

Butler, C., Parkhill, K. A., Shirani, F., Henwood, K., & Pidgeon, N. (2014). Exploring the dynamics of energy demand through a biographical lens. *Nature and Culture, 9*(2), 164–182.

Cox, E., Royston, S., & Selby, J. (2019). From exports to exercise: How non-energy policies affect energy systems. *Energy Research & Social Science, 55*, 179–188.

Day, R., Walker, G., & Simcock, N. (2016). Conceptualising energy use and energy poverty using a capabilities framework. *Energy Policy, 93*, 255–264.

Foucault, M. (1991). Governmentality. In G. Burchell, C. Gordon, & P. Miller (Eds.), *The Foucault effect: Studies in governmentality* (pp. 87–105). The University of Chicago Press.

Gormally, A. M., O'Neill, K., Hazas, M. D., Bates, O. E. G., & Friday, A. J. (2019). "Doing good science": The impact of invisible energy policies on laboratory energy demand in higher education. *Energy Research & Social Science, 52*, 123–131.

Greene, M., & Fahy, F. (2020). Steering demand? Exploring the intersection of policy, practice and lives in energy systems change in Ireland. *Energy Research & Social Science, 61*, 101331.

Middlemiss, L. (2016). A critical analysis of the new politics of fuel poverty in England. *Critical Social Policy, 37*(3), 425–443.
Middlemiss, L., Ambrosio-Albalá, P., Emmel, N., Gillard, R., Gilbertson, J., Hargreaves, T., Mullen, C., Ryan, T., Snell, C., & Tod, A. (2019). Energy poverty and social relations: A capabilities approach. *Energy Research & Social Science, 55*, 227–235.
Middlemiss, L., & Gillard, R. (2015). Fuel poverty from the bottom-up: Characterising household energy vulnerability through the lived experience of the fuel poor. *Energy Research & Social Science, 6*, 146–154.
Morley, J., Widdicks, K., & Hazas, M. (2018). Digitalisation, energy and data demand: The impact of internet traffic on overall and peak electricity consumption. *Energy Research & Social Science, 38*, 128–137.
Nussbaum, M. C. (2011). *Creating capabilities: The human development approach*. Harvard University Press.
Petrova, S. (2018). Encountering energy precarity: Geographies of fuel poverty among young adults in the UK. *Transactions of the Institute of British Geographers, 43*(1), 17–30.
Royston, S. (2016). *Invisible energy policy in higher education*. Conference paper at the DEMAND Centre Conference April 2016, Lancaster. Available at: https://www.demand.ac.uk/wp-content/uploads/2016/05/Royston-Invisible-energy-policy-in-HE.pdf
Royston, S., Selby, J., & Shove, E. (2018). Invisible energy policies: A new agenda for energy demand reduction. *Energy Policy, 123*, 127–135.
Sen, A. (1999). *Development as freedom*. Alfred Knopf.
Shove, E. (2002, October). *Rushing around: Coordination, mobility and inequality*. Draft paper for the Mobile Network meeting.
Shove, E. (2003). Converging conventions of comfort, cleanliness and convenience. *Journal of Consumer Policy, 26*(4), 395–418.
Shove, E. (2004). Efficiency and consumption: Technology and practice. *Energy and Environment, 15*(6), 1053–1066.
Shove, E. (2010). Beyond the ABC: Climate change policy and theories of social change. *Environment and Planning A: Economy and Space, 42*(6), 1273–1285.
Shove, E. (2017). What is wrong with energy efficiency? *Building Research & Information, 46*(7), 779–789.
Shove, E. Pantzar, M., & Watson, M. (2012). *The dynamics of social practice: Everyday life and how it changes*. Sage.
Simcock, N., Walker, G., & Day, R. (2016). Fuel poverty in the UK: Beyond heating? *People Place and Policy Online, 10*(1), 25–41.
Walker, G. (2013). Inequality, sustainability, and capability: Locating justice in social practice. In E. Shove & N. Spurling (Eds.), *Sustainable practices: Social theory and climate change* (pp. 181–196). Routledge.

CHAPTER 2

Poverty and Energy Demand

Abstract The starting point for this book is to explicate key conceptual interventions within the large body of existing work on energy poverty. This chapter discusses the trajectories of this research tradition from analyses focused more narrowly on notions of fuel poverty to those foregrounding issues of energy vulnerability, and through to concepts of precarity and capabilities. It examines the frontiers of analysis engaging with work that has sought to move beyond preoccupations with heat and the spatial domain of the home and advance more complex understandings of the issues. The chapter concludes setting out the key tenets of the literature and discussing how conceptualisations can be further advanced in analysis of energy poverty.

Keywords Fuel poverty · Energy poverty · Energy vulnerability · Capabilities · Precarity

Introduction

Brenda Broadman (1991) is often credited with bringing prominence to fuel poverty as a focus for academic analysis. Her work underpinned the formulation of a key definition for fuel poverty within the UK. This definition positioned fuel poverty as arising in contexts where a household spends more than 10% of their income to afford adequate domestic

C. Butler, *Energy Poverty, Practice, and Policy*, Progressive Energy Policy,
https://doi.org/10.1007/978-3-030-99432-7_2

energy services, particularly heat (though electricity for other energy uses such as lighting was also incorporated). An understanding of fuel poverty as related primarily to the ability of households to heat their homes to an adequate standard has since been pervasive across both academic and policy analysis (Simcock et al., 2016). In application, particularly in the UK, this focus has combined with a tendency to characterise fuel poverty as an issue experienced by older people (Day & Hitchings, 2011; Simcock et al., 2016). Such approaches to the analysis of fuel poverty have, however, been critiqued for failing to engage with lived experiences and underlying systemic causes, as well as for working from a narrow definition of the relevant energy uses to be considered (e.g. Bouzarovski & Petrova, 2015; Day et al., 2016; Gillard et al., 2017; Middlemiss & Gillard, 2015; Petrova, 2018; Simcock et al., 2016; Walker et al., 2016).

Academic debates about fuel poverty have advanced taking in the wider subfield of energy justice and introducing new concepts that offer a deeper basis for engagement with these issues (e.g. Bouzarovski & Petrova, 2015; Day & Walker, 2013; Day et al., 2016). In particular, the concepts of energy vulnerability, precarity, and capabilities have gained traction and discussion has moved on from a focus on older people, heat, and costs versus income towards consideration of the multifaceted nature of fuel poverty (e.g. Bouzarvoski, 2018; Bouzarovksi & Petrova, 2015; Gillard et al., 2017; Middlemiss & Gillard, 2015; Petrova, 2018; Simcock et al., 2016). These multidimensional understandings have also sought to traverse a traditional spatial focus on domestic social contexts, looking at interconnections and relationality beyond the home (Bouzarovski & Petrova, 2015; Day et al., 2016; Middlemiss et al., 2019). However, though mobilities are raised within this work as an area of energy poverty (e.g. Middlemiss et al., 2019), analyses have not yet gone as far as to connect with the burgeoning literature on transport poverty (e.g. Mattioli, 2017; Robinson & Mattioli, 2020). This literature offers important insights regarding the negative impacts of transport poverty in terms of wellbeing, hardship, and social exclusion, as well as advancing understanding of the drivers.

Scholarship building from a concern with transport poverty has also sought to make connections with the domestic energy poverty literature. Robinson and Mattioli (2020) take forward an extensive quantitative analysis of the potential within England for 'double energy vulnerability'—a situation whereby a household is doubly vulnerable to both domestic and transport poverty. However, the analyses in this area tend to be primarily substantive in focus engaging far less with the conceptual debates that have characterised the domestic energy poverty field.

This chapter discusses these key debates, concerns, and conceptual advances that characterise contemporary energy poverty research. In this, it brings together substantive insights from across fuel and transport poverty, conceptual concerns spanning ideas of vulnerabilities, precarity, and capabilities, and key issues that have been raised through engagement with lived experiences. Through discussion of these different contributions, I seek to advance conceptual understanding and set out an approach to energy poverty that encompasses the multiple dimensions, concerns, and spaces across this wide-ranging literature.

From Fuel Poverty to Energy Vulnerability and Precarity

Fuel poverty has gained prominence in academic research and policy since the 1980s but in recent years it has come to the fore as a prominent global issue that has been allied with problems of energy access (Bouzarovski & Petrova, 2015). Issues related to energy and poverty have come to be recognised within academic research as multifaceted phenomena that are constituted through interaction between multiple personal, social, economic, and political dimensions. Bouzarovski et al. (2021) have highlighted how energy poverty research has expanded well beyond concerns with low incomes, high energy prices, and residential energy efficiency. Instead, recent research trajectories have focused on a wide range of factors that produce *vulnerability* to energy poverty, as well as emphasising differences in lived experience across people and place (e.g. Butler & Sherriff, 2017; Gillard et al., 2017; Middlemiss & Gillard, 2015; Petrova & Simcock, 2019).

As an approach, 'energy vulnerabilities' have been argued to 'help draw a distinction between energy or fuel poverty as a descriptor of a state within a certain temporal frame, on the one hand, and vulnerability as a set of conditions leading to such circumstances, on the other' (Bouzarovski, 2018, p. 18). Indeed, Middlemiss and Gillard (2015, p. 147) have defined energy vulnerability as: 'the likelihood of a household being subject to fuel poverty, the sensitivity of that household to fuel poverty, and the capacity that household has to adapt to changes in fuel poverty'. But they also highlight how this is 'somewhat unsatisfactory' given its failure to engage with the complexity of lived experience. They discuss some of the critiques of 'vulnerability' and the linked concept of 'resilience'. They highlight arguments concerning the tendency of these concepts to depoliticise vulnerability challenges by individualising and displacing responsibility away from the wider social and political sphere. For Middlemiss and Gillard (2015), however, this does not mean vulnerability must be abandoned altogether but, rather, that it requires recognition of dimensions of power when used.

One of the departure points for energy vulnerability research, then, is the realisation that fuel poverty is not a static condition but is better understood in terms of the factors that might cause its emergence (Bouzarovski, 2018; Day & Walker, 2013; Meyer et al., 2018; Middlemiss & Gillard, 2015). Such factors that contribute to its emergence in domestic homes have been described as: quality of dwelling fabric; tenancy relations; energy costs and supply; stability of household income; social relations in and out of the household; and ill health (Middlemiss & Gillard, 2015, p. 149). Added to this are institutional and socio-political factors that shape vulnerabilities across different geographic contexts (Bouzarovski et al., 2015; Petrova & Prodromidou, 2019; Bouzarovski et al., 2021). This positions the responses to fuel poverty as ones that involve addressing the wider underlying factors that can see people move in or out of fuel poverty across different times and spaces (Bouzarovski et al., 2021; Middlemiss & Gillard, 2015). Here a focus on precarity has been used in efforts to characterise energy poverty and deepen the concept of energy vulnerability.

In particular, Petrova (2018) has given the concept of precarity a central place in understanding the constitution of energy vulnerability. She argues that while *precariousness* is a shared condition related, for example, to a particular sector such as housing, energy *precarity* is a politically induced phenomenon that is only generated under certain circumstances. Petrova explains that the concept of precarity, widely explored in other literature outside of energy research, has roots in Bourdieu's practice theoretical analysis and characterises people that are both vulnerable and marginalised but also transformative as a class of people in-the-making. In this sense, she highlights how the concept of precarity overcomes some of the critiques associated with 'vulnerability' and the related concept of resilience by positioning it as a politically induced phenomenon that also affords 'agency for political change and emancipation' (2018, p. 19). As such Petrova (2018, p. 20) argues that 'energy precarity' can be used as:

> a double signifier that calls attention to the performative experience of multiple vulnerabilities in the home, while illuminating the political and institutional embeddedness of fuel poverty.

In this way, she proposes energy precarity as a complementary concept to those of fuel poverty and energy vulnerability suggesting that it expands 'understandings of energy deprivation beyond the home, and [links] them with the institutional and political circumstances that may im/mobilise particular socio-demographic groups to act on the issue' (Petrova, 2018, p. 20). Here, attention is brought onto the conditioning of precarity through wider social, political, and economic dynamics. For example, Petrova highlights how within the UK's private-rented and houses in multiple occupation (HMO) sectors, a lack of strategic governance and reliance on voluntary improvements by landlords has contributed to the dominance of poorly insulated and old homes. She further argues that the short-term and transient nature of much of the occupancy of private-rented accommodation in the UK disincentivises landlords from improving housing, as well as contributing to a normalisation of energy deprivation amongst young people. Petrova explains 'accepting the mainstream framing of poor living conditions as provisional and non-permanent made living in fuel poverty tolerable for the interviewees' (2018, p. 26). Ultimately, Petrova shows how these socially and politically constituted trends in the UK private-rented sector shape experiences of energy deprivation.

The concepts of energy vulnerability and (following Petrova) precarity thus allow for a stronger characterisation of the variability of circumstances and processes through which experiences of energy deprivation are made manifest. Precarity, however, arguably has greater potential to move beyond the often neoliberal, individualised characterisations of energy deprivation that have tended to pervade the concept of vulnerability. Despite this, both energy vulnerability and precarity have proven useful for thinking outside of the preoccupations of particular contexts (such as domestic settings) and engaging with inequalities relating to energy use in less constrained ways, such as those dictated by frequently narrow policy definitions.

This is borne out by analyses that have specifically sought to think about energy poverty issues more expansively, moving past the conventional focus on particular energy uses (e.g. heat) and specific demographics (e.g. older people). In this vein, Simcock et al. (2016) bring focus on non-heat related energy uses, giving emphasis to domestic *electricity* using energy services. Looking at the UK case, their analysis shows how despite inclusion of non-heat home energy uses within policy (e.g. electric appliance use), there remains a strong emphasis on heat as the main focus for policy and governance responses. They assert that there is 'significant scope for further investigation… on how and why vulnerable households may suffer "under -consumption" in *non-heating* energy-uses, and moreover on how this impacts upon different dimensions of people's quality of life' (2016, p. 37).

They also highlight the need for more forward-looking analyses of energy vulnerabilities that account for wider societal changes and shifts in the nature of basic necessities, highlighting consumer electronics as an important area for research (Simcock et al., 2016). Petrova (2018) similarly emphasises the importance of examining non-heat energy services in her work on the experiences of younger people; a demographic not typically addressed by energy poverty research. She shows the heightened importance of energy services connected to information and communication technologies within the lives of younger people, highlighting the significance of looking beyond both heat and older people in energy poverty research. Beyond this, a small but important body of work has brought focus on low income and disabled people (e.g. Gillard et al., 2017; Snell et al., 2015) as groups that are inadequately recognised and addressed through existing fuel poverty policy.

The shift to look beyond the prior preoccupations of energy poverty research and policy encompassing multiple dimensions, varied energy services, and different demographic groups has been accompanied by calls to move outside a focus on the spatial context of the home. This has taken different forms with some arguing for greater attentiveness to the ways that domestic energy deprivation is negotiated and constituted beyond the confines of the home (Bouzarovski & Petrova, 2015; Petrova, 2018). For example, Petrova (2018) highlights how the young people in her study often deployed strategies of visiting others or spending time in other spaces to which they had access, like offices or libraries, to mitigate their energy deprivation. Others have led calls to look at the intersections between domestic energy poverty and transport poverty (e.g. Robinson & Mattioli, 2020), seeking to bring focus on the relevance of mobilities for energy poverty research. Here, analysis highlights the lack of research addressing the intersections between transport poverty and domestic energy poverty. Though the emergence of more complex spatially integrated understandings offers significant potential to open-up analysis to mobilities, such integration has yet to form a focus for much empirical analysis within energy poverty research. The transport poverty literature has, however, begun to develop analyses that seek to bridge the divide, and it is to discussion of this the following section turns.

Bringing Transport Poverty and Mobilities into Focus

As highlighted above, an important precursor to work that seeks to draw mobilities into the wider domain of energy poverty research is found in the transport poverty literature. The body of literature around transport poverty, with links to work on mobility and social exclusion (e.g. Cass et al., 2005; Currie, 2011), emerged initially largely in isolation from work on domestic energy vulnerability and fuel poverty (Robinson & Mattioli, 2020). This research tradition has long addressed issues of car access, transport affordability, costs of motoring, and vulnerability to fuel price increases, with issues of 'forced car ownership' shaping research agendas in this space (e.g. Currie, 2011; Currie et al., 2007; Mattioli, 2017; Mattioli et al., 2017). Analysis of socio-spatial configurations has also formed a focus bringing into view the relations between income, place of dwelling, accessibility, and vulnerability to transport poverty (e.g. Curl et al., 2011; Mattioli, 2017). However, the work in this space has rarely engaged with the conceptual debates that characterise some of the wider energy poverty literature. It has developed instead focusing on concepts of social exclusion, accessibility, and more recently justice, producing distinctive insights.

For instance, Mullen and Marsden (2016) have used a justice conceptual lens to highlight the longstanding set of processes that have favoured private car travel as the primary mode of transport within the UK as pertinent to social exclusion. They argue that mobility systems that privilege those who can access a private vehicle and afford to use it raises important justice concerns as it inhibits the welfare of those for whom it is not possible (e.g. those without economic means to support car use). This suggests the importance of examining how demands for specific forms of travel are constituted over time, something that aligns with practice-oriented thinking about mobilities (discussed in the next chapter).

Though there is little work at the intersection of domestic and transport poverty, there are some notable contributions that have sought to bridge the divide. For example, Mattioli et al. (2017) offer an important comparison of fuel poverty and transport poverty. They argue that while transport affordability problems have typically been based on an analogy with the more dominant issue of fuel poverty, important conceptual differences between the two issues can be identified. For example, they discuss research that has shown how people are more likely to prioritise transport over other energy costs (such as heating), because of their requirements for work. This is suggestive of an important recursive link between employment and economic stress related to transport that they argue has no clear parallel in the context of fuel poverty (Mattioli et al., 2017). They assert the importance of examining the interaction between different dimensions of energy poverty (i.e. between fuel and transport poverty), given the connections and differences between them that have been identified. More broadly, they emphasise how the focus in most transport poverty research has been on car-dependency, rather than looking at mobilities and a diverse range of modal forms. And finally, they reflect a need, highlighted above (Petrova, 2018; Simcock et al., 2016), to examine the ways these issues affect different demographic groups beyond the elderly, given what they assert to be an overemphasis in policy on universal measures for older people.

In other work, Robinson and Mattioli (2020) have developed an analysis to show the relatively widespread occurrence of what they term 'double energy vulnerability' that arises where vulnerabilities to domestic energy poverty intersect with vulnerabilities to transport poverty. Developing a high-level spatial analysis, they look at the overlaps using quantitative indicators of vulnerability to transport and domestic energy poverty. They show that as many as 6% of neighbourhoods accounting for 3 million residents have a high propensity towards double energy vulnerability (Robinson & Mattioli, 2020). Though this work offers insights into the connections between transport and fuel poverty, I argue that it is important to go further and engage with mobilities as part of a broader conceptualisation. I propose that the conceptual advances in work developing vulnerability, precarity, and capabilities approaches can provide fertile ground for this. It is to a discussion of the interventions in energy poverty research developing capabilities-based analyses that I now turn.

The Capabilities Approach in Energy Poverty Research

A key intervention within the energy poverty literature has been to develop and apply a capabilities-based approach to understanding energy deprivation (Day et al., 2016; Middlemiss et al., 2019). This has built on existing work to facilitate understandings of energy poverty as a complex, multifaceted phenomenon, while also bringing focus on energy services and the connections between energy and human needs. As an approach to energy poverty, it has important appeal for the potential it holds to engage with the complexities of lived experiences and the manifold intersections that shape them.

The energy capabilities approach is predicated on Sen and Nussbaum's understanding of human wellbeing as requiring certain capabilities to support opportunities for functioning (see Sen, 1999; Nussbaum, 2003, 2011; Sen & Nussbaum, 1993). They argue that the focus for societal development should be on key basic functionings for human wellbeing, such as bodily health. They maintain, however, that there are multiple capabilities, such as the ability to secure healthy food, that underpin these basic functionings, and it is these capabilities that must be the goal because functionings can be an outcome of choice. For example, a person who has the means to secure food but chooses not to eat might have the same level of functioning as a person who does not have sufficient access to food, but these are clearly not the same.

Day et al. (2016), in particular, have applied this line of theorisation about human wellbeing to energy deprivation putting forward a framework for examining energy poverty. They draw on Smith and Seward's (2009) distinction between *basic capabilities*, such as maintaining good health, having social respect or being educated, and *secondary capabilities* that underpin basic capabilities, such as washing clothes or storing and preparing food or accessing information and resources. Day et al. argue that many of these secondary capabilities often require energy in some form and relate therefore to different energy services (see Fig. 2.1 from Day et al., 2016, p. 260)

They further highlight how the energy needs required to fulfil different capabilities are shaped by particular characteristics and circumstances (e.g. whether you are young, old, disabled, healthy or ill), material factors (e.g. the type of home you live in, the local climate), and the availability of

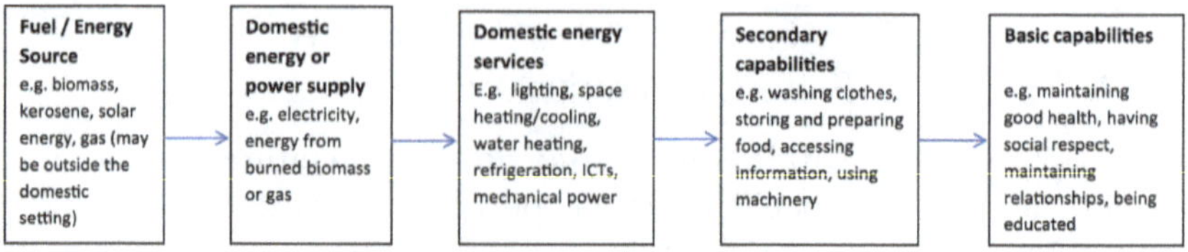

Fig. 2.1 Conceptualising the relationship between energy, services, and outcomes

energy services beyond the home (such as through wider societal infrastructure and community services). This understanding, thus, connects to ideas of energy vulnerability in denoting the characteristics, circumstances, and materials that shape experiences of energy poverty.

They add to this an explanation of how societal norms, such as those pertaining to cleanliness, shape energy service needs in terms of how they relate to the basic and secondary capabilities. For example, in the UK showering every day has become a normal expectation for most people to ensure social respect is maintained. This means that many secondary capabilities, which for Day et al. underpin basic capabilities, have variable implications for energy services and resource use depending on the specifics of place and the prevailing social norms. They arrive at a distinctive and flexible definition of energy poverty that is as follows:

> an inability to realise essential capabilities as a direct or indirect result of insufficient access to affordable, reliable and safe energy services, and taking into account available reasonable alternative means of realising these capabilities. (Day et al., 2016, p. 260).

There are several implications of this alternative definition. First, it offers a multidimensional approach that is closer to understandings of energy poverty typically used in Global South contexts, which recognise the importance of energy for capabilities and wellbeing (Bouzarovski & Petrova, 2015; Day et al., 2016). It thus allows for a greater ability to perceive the complex interdependencies between energy and poverty (Bouzarovski, 2018). Second, it encompasses key assertions from energy vulnerability research by highlighting how different characteristics shape energy needs, as well as abilities to meet them. Third, it recognises the role of energy services but does not specify services giving the required flexibility to engage with different cultural and material contexts and societal norms (Day et al., 2016).

Fourth, it offers a means of making visible 'the effect of evolving social norms in constituting energy demand and, therefore, relative energy deprivation' and explicitly identifying such processes as spaces of intervention for alleviation of fuel poverty (2016, p. 262). Day et al. assert that 'energy poverty can involve not being able to engage in accepted social practices' (ibid.) and suggest interventions designed to reduce demand for energy services might be as relevant to alleviating energy poverty as they are to sustainability. For example, it brings into view questions about how capabilities might be supported in ways other than increasing the amount of energy required (e.g. building design that incorporates cooling, rather than air conditioning), and how there can be different ways of providing services beyond an individualised focus (e.g. through community-based provision of ICT infrastructure).

Such understandings are foreshadowed in debates about 'energy services' as a basis for characterising energy poverty (Bouzarovski & Petrova, 2015). Here, the emphasis is similarly on the functions that energy affords and the abilities that people have for achieving those functions to a satisfactory level. Thinking in this way lends itself to consideration of the wider technologies and dynamics involved in fulfilling energy services and thus implicated in experiences of energy poverty. Bouzarovski and Petrova (2015, p. 34) highlight how 'while studies of consumption and sustainability have often explored the ways that particular patterns of energy use are normalised via social practices and everyday routines, there has been little work on the levels of domestic energy services that households require for full participation in society' (Bouzarovski & Petrova, 2015, p. 34). They argue that such an emphasis on required service levels that allow for opportunities to undertake actions and activities offers an important starting point for advancing global efforts to understand and address energy poverty. This suggests the importance of looking at patterns of energy use within energy poverty research, but it stops short of questioning the processes through which needs are constituted.

The capabilities approach, and related analyses, thus offer a great deal in terms of facilitating understanding of energy poverty that moves away from the modelled measures and definitions characteristic of current policy (see Chapter 4 for discussion). However, I aim to develop these conceptual ideas further by engaging with some of the wider energy poverty literature discussed thus far, as well as some existing critical engagement (Middlemiss et al., 2019). This is addressed in the final section of this chapter.

Advancing an Approach to Energy Poverty

A useful mechanism for explicating the expanded capabilities approach to energy poverty proposed here is to work through the ways that the conceptualisation differs from Day et al.'s (2016) intervention. In discussing key areas of difference, other concepts and analytic ideas are brought in to advance an approach to energy poverty that synthesises and solidifies major aspects of thinking in this space.

First, in a similar way to others working with wider perspectives on energy poverty, Day et al. (2016) do not explicitly use the flexibility inherent in their conceptualisation to look across to mobilities. They focus on largely domestic energy services (e.g. heating, lighting), though they do engage with forms of service provisioning beyond the home. For example, they suggest that their definition gives room for energy services provisioning, such as washing, to be outside of the home via more communal modes of delivery (Day et al., 2016). The intention here, however, and one of the appeals of this approach to defining energy poverty, is that it can take in deprivation associated with travel and mobilities, as well as those related to domestic contexts. Given that mobilities have received little attention in the debates about defining energy poverty, it seems a missed opportunity not to use the capabilities approach to further open-up the conception. For this reason too, mobilities form a focus for key parts of the empirical analysis that comes in later chapters of this book.

A second point of distinction taken forward within the conceptualisation advanced here relates to arguments that Day et al. (2016) make about the need to distinguish between capabilities that are understood as essential, and those which are not, across different contexts and places. They suggest that there might be a need to decide 'threshold levels' for some basic and secondary capabilities (2016, p. 261). In discussing how this might be achieved, they refer to the wider capabilities literature and suggest two routes to identifying essential capabilities; to work from a list of capabilities (similar to that developed by Nussbaum) or to develop understanding of essential capabilities and their relationship to energy through some form of deliberative process within particular contexts (as advocated by Sen). Day et al. (2016) favour situated deliberation and grounded analyses of energy poverty as the best route to identifying essential capabilities.

However, while universal definitions (as developed by Nussbaum) have been challenged for lacking insight into important place-based differences, so too have relative characterisations that rely on deliberation as recommended by Sen. Indeed, this latter approach has been critiqued for obscuring relations of oppression and marginalisation that shape what people view as normal and acceptable in terms of human needs (Deneulin & McGregor, 2010; Mahali et al., 2018). Here, it is useful to reflect on Petrova's (2018) exploration of precarity as a concept that brings into view broader socio-political contexts and their implications for people's conceptions of fuel poverty.

It is possible to see how the argument that conceptions of need are shaped by socio-political conditions relates strongly to ideas of precarity that foreground the 'socio-institutional normalisation' of (energy) poverty in ways that 'immobilise' people from becoming politically active (Petrova, 2018, p. 18). In this context, while Day et al. (2016) suggest deliberation as a route to defining capabilities, I argue that such attempts to specify essential capabilities and their related energy requirements could do more to obscure connections and interrelations than they reveal. This is because, as Petrova's (2018) work demonstrates, people living without proper access to energy services to support basic capabilities do not necessarily problematise this deficit precisely because of processes of socio-political normalisation. This suggests a challenge to notions that the most fruitful way of taking forward a capabilities-based analysis of energy poverty lays in specifying capabilities and the links to energy.

For this reason, though the capabilities and energy services that are of interest for thinking about energy poverty must be discussed through analysis, I argue that it is not necessary (or perhaps desirable) to produce a comprehensive list of any sort. Indeed, I would suggest that the focus of analysis should not be on producing insight into the *range* of essential capabilities implicated in or affected by energy poverty as this would only serve to constrain and delimit the more flexible understanding of energy poverty that the capabilities framework affords. Instead, a capabilities-based approach could more fruitfully be used to facilitate an alertness to both a wide range of energy services beyond those that dominate debates *and* the interconnections between them and capabilities. Given this, in the later chapters of this book, the energy services addressed through the analysis are in no way exhaustive and do not address all areas of need. Rather, I select examples for the insights they give into the links between capabilities and energy services and for their relevance to understanding the dynamics of energy poverty.

The third area of difference concerns the different proposals for how to approach researching and analysing energy poverty using these wider conceptualisations. Where other analyses suggest a focus on energy services, needs, or the capabilities at issue, I adopt a biographical approach (see Butler et al., 2014) that places the person and their relational context at the centre. This prevents interconnections between different forms of energy poverty from being obscured and opens-up the analysis to complexity by engaging with lived experiences. I suggest it can facilitate movement past the traditional spatial boundedness of energy poverty research and offer a route to engaging with capabilities in a grounded way without the need to specify essential needs.

A fourth and final point concerns the way the relationship between capabilities and energy services has been depicted within energy capabilities work to date. Middlemiss et al. (2019) argue that Day et al. (2016) effectively suggest a sequential relationship between domestic energy services, secondary capabilities, and basic capabilities. This, they suggest, implies that basic capabilities are in effect served by secondary capabilities and the related energy services, not the other way around. Middlemiss et al.'s (2019) intervention focuses on social relations as a basic capability, highlighting a more bidirectional relationship than this conception affords. They argue that social relations cannot be adequately characterised as either 'secondary' or 'basic' capabilities as they 'might be both an end in themselves… as well as a means by which other ends could be achieved' (Middlemiss et al., 2019, p. 229). This brings into question the value of distinguishing between secondary and basic capabilities or characterising the direction of the relationship between them and energy services.

Given this, while the existing conceptual discussions of energy capabilities and services have been characterised by frameworks and schematics that offer means for navigating the complexities inherent in energy poverty (e.g. see Bouzarovski & Petrova, 2015; Day et al., 2016), I wish to advocate a movement away from specifications of this nature instead opting to embrace the complexity and unravel it through and within empirical analysis. Again, a biographical methodology can facilitate this as suggested above, but other methodologies could be applied with the same orientation. The important point is to maintain openness to the complexities inherent in the relations between energy services and capabilities.

This could be in the ways suggested by Middlemiss et al. (2019), but it is possible to think of other forms of complexity, such as in questioning what it means to have a capability and its relation to energy services. For example, the capability to shower can involve a 2-minute cold shower or a 20-minute hot shower—while both might be regarded as having capabilities to meet a basic need of cleanliness, the former does not achieve the same level of functioning as the latter. Understanding of self-rationing and self-disconnection also calls into question assumptions about 'access' to energy in terms of infrastructural provisioning, since it cannot be assumed that the capability automatically follows from availability of the service. This variability in the relations between energy services and capabilities means that it is extremely difficult—and I argue potentially not desirable—to apply either a broader top-down or a very closely specified approach to analysis of the relations between energy services and capabilities. Instead, it is possible to keep the contours of need and the *extent* to which needs are being met or not as an integral and emergent part of analysis, rather than attempting to develop and apply distinctions.

In sum, the approach to energy poverty advocated here is one that encompasses an understanding of energy in terms of *what it is for*, focusing on energy services and related capabilities. But it also extends beyond the preoccupations of capabilities and wellbeing research more generally in not seeking to provide a list of all relevant capabilities or detail specific connections to energy services. Rather, the approach is one that calls for focus on the situated and relational person and their experiences to keep complexity and interconnections in view.

Conclusion

This chapter has discussed some of the major contemporary contributions to conceptualising energy poverty and argued for their value in instituting an approach that can keep complexity in view. The centrality of the capabilities approach, along with the related concept of energy services, has been explicated and explored. I arrive at a characterisation of energy poverty through the lens of lived experiences that reflects a distinctive understanding of the issues, departing markedly from contemporary policy definitions. Such an understanding entails a focus on the impacts of lack of access not just to energy but to the capabilities that energy use is implicated in, including being healthy, educated, and able to socially participate. Included in this is an ability to reflect on how issues of energy poverty are bound up not only with the point of energy use but with the materials, wider infrastructure, and social processes that shape energy services.

The understanding of energy deprivation advanced here, then, is one that affords space to different energy services bringing into view areas of interconnection and complexity in how energy use is prioritised and negotiated as part of efforts to live and fulfil basic functionings. In this, I argue it is paramount to understand energy deprivation in terms of what happens both within and beyond the home. For Day et al. (2016), such an unbounding of energy deprivation facilitates engagement with spaces beyond the home focusing on more communal forms of energy service. Elsewhere this approach has revealed how *strategies for coping* with fuel poverty often extend spatially too (Petrova, 2018). In this latter context, uses of spaces beyond the home as ways to meet needs for energy services have been cast critically. For example, Petrova (2018, p. 24) highlights how tendencies to use spaces outside of the home for warmth as a way of dealing with cold homes can contribute to 'the intensification and normalisation of [energy] precarity'. However, despite the inherent possibilities very little research has yet moved to analysis that draws in mobilities as well. I argue, here, this opening up is afforded by approaches grounded in capabilities and precarity and could be advanced much more strongly within future analysis.

The chapter has foregrounded the relevance of combining the multi-faceted approach to energy poverty encapsulated in the capabilities framework (Bouzarovski & Petrova, 2015; Day et al., 2016; Middlemiss et al., 2019) with the concerns and complexities that Petrova (2018) brings into view by mobilising precarity. Crucially, for the intentions of this book, both approaches align with and link to practice theory. However, I argue there is also much more to be gained from a deeper and more explicit engagement with practice theory in concert with these approaches. For example, fruitful avenues for analysis can be found in the ways that practice theory research brings far greater focus on how energy needs are actively constituted by policies, processes, and interventions. In the following chapter, I turn to the practice theory literature on energy demand introducing the key tenets of this work that are important for the analysis in this book.

References

Boardman, B. (1991). *Fuel poverty: From cold homes to affordable warmth*. Belhaven Press.

Bouzarovski, S. (2018). *Energy poverty: (Dis)assembling Europe's infrastructural divide*. Palgrave.

Bouzarovski, S., & Petrova, S. (2015). A global perspective on domestic energy deprivation: Overcoming the energy poverty–fuel poverty binary. *Energy Research & Social Science, 10*, 31–40.

Bouzarovski, S., Thomson, H., & Cornelis, M. (2021). Confronting energy poverty in Europe: A research and policy agenda. *Energies, 14*(4), 858.

Bouzarovski, S., Tirado Herrero, S., Petrova, S., & Ürge-Vorsatz, D. (2015). Unpacking the spaces and politics of energy poverty: Path-dependencies, deprivation and fuel switching in post-communist Hungary. *Local Environment, 21*(9), 1151–1170.

Butler, C., Parkhill, K. A., Shirani, F., Henwood, K., & Pidgeon, N. (2014). Exploring the dynamics of energy demand through a biographical lens. *Nature and Culture, 9*(2), 164–182.

Butler, D., & Sherriff, G. (2017). "It's normal to have damp": Using a qualitative psychological approach to analyse the lived experience of energy vulnerability among young adult households. *Indoor and Built Environment, 26*(7), 964–979.

Cass, N., Shove, E., & Urry, J. (2005). Social Exclusion, mobility and access. *The Sociological Review, 53*(3), 539–555.

Curl, A., Nelson, J. D., & Anable, J. (2011). Does accessibility planning address what matters? A review of current practice and practitioner perspectives. *Research in Transportation Business & Management, 2*, 3–11.

Currie, G. (Ed.). (2011). *New perspectives and methods in transport and social exclusion research*. Emerald Group Publishing Limited.

Currie, G., Stanley, J., & Stanley, J. (2007). *No way to go: Transport and social disadvantage in australian communities*. Monash University Press.

Day, R., & Hitchings, R. (2011). "Only old ladies would do that": Age stigma and older people's strategies for dealing with winter cold. *Health & Place, 17*(4), 885–894.

Day, R., & Walker, G. (2013). Household energy vulnerability as assemblage. In K. Bickerstaff, G. Walker, & H. Bulkeley (Eds.), *Energy justice in a changing climate* (pp. 14–29). Zed Books.

Day, R., Walker, G., & Simcock, N. (2016). Conceptualising energy use and energy poverty using a capabilities framework. *Energy Policy, 93*, 255–264.

Deneulin, S., & McGregor, J. A. (2010). The capability approach and the politics of a social conception of wellbeing, *European Journal of Social Theory, 13*(4), 501–519

Gillard, R., Snell, C., & Bevan, M. (2017). Advancing an energy justice perspective of fuel poverty: Household vulnerability and domestic retrofit policy in the United Kingdom. *Energy Research & Social Science, 29*, 53–61.

Mahali, A., Lynch, I., Fadiji, A. W., Tolla, T., Khumalo, S., & Naicker, S. (2018). Networks of well-being in the global south: A critical review of current scholarship. *Journal of Developing Societies, 34*(4), 373–400.

Mattioli, G. (2017). "Forced car ownership" in the UK and Germany: Socio-spatial patterns and potential economic stress impacts. *Social Inclusion, 5*(4), 147.

Mattioli, G., Lucas, K., & Marsden, G. (2017). Transport poverty and fuel poverty in the UK: From analogy to comparison. *Transport Policy., 59*, 93–105.

Meyer, S., Laurence, H., Bart, D., Middlemiss, L., & Maréchal, K. (2018). Capturing the multifaceted nature of energy poverty: Lessons from Belgium. *Energy Research & Social Science, 40*, 273–283.

Middlemiss, L., & Gillard, R. (2015). Fuel poverty from the bottom-up: Characterising household energy vulnerability through the lived experience of the fuel poor. *Energy Research & Social Science, 6*, 146–154.

Middlemiss, L., Ambrosio-Albalá, P., Emmel, N., Gillard, R., Gilbertson, J., Hargreaves, T., Mullen, C., Ryan, T., Snell, C., & Tod, A. (2019). Energy poverty and social relations: A capabilities approach. *Energy Research & Social Science, 55*, 227–235.

Mullen, C., & Marsden, G. (2016). Mobility justice in low carbon energy transitions. *Energy Research & Social Science, 18*, 109–117.

Nussbaum, M. C. (2003). Capabilities as fundamental entitlements: Sen and global justice. *Feminist Economics, 9*(2–3), 33–59.

Nussbaum, M. C. (2011). *Creating capabilities: The human development approach*. Harvard University Press.

Petrova, S. (2018). Encountering energy precarity: Geographies of fuel poverty among young adults in the UK. *Transactions of the Institute of British Geographers, 43*(1), 17–30.

Petrova, S., & Prodromidou, A. (2019). Everyday politics of austerity: Infrastructure and vulnerability in times of crisis. *Environment and Planning C: Politics and Space, 37*(8), 1380–1399.

Petrova, S., & Simcock, N. (2019). Gender and energy: Domestic inequities reconsidered. *Social & Cultural Geography, 22*(6), 849–867.

Robinson, C., & Mattioli, G. (2020). Double energy vulnerability: Spatial intersections of domestic and transport energy poverty in England. *Energy Research & Social Science, 70*, 101699.

Sen, A. (1999). *Development as freedom*. Alfred Knopf.

Sen, A., & Nussbaum, M. (Eds.). (1993). *The quality of life*. Oxford University Press.

Simcock, N., Walker, G., & Day, R. (2016). Fuel poverty in the UK: Beyond heating? *People Place and Policy Online, 10*(1), 25–41.
Smith, M. L., & Seward, C. (2009). The relational ontology of Amartya Sen's capability approach: Incorporating social and individual causes. *Journal of Human Development and Capabilities, 10*(2), 213–235.
Snell, C., Bevan, M., & Thomson, H. (2015). Welfare reform, disabled people and fuel poverty. *Journal of Poverty and Social Justice, 23*(3), 229–244.
Walker, G., Simcock, N., & Day, R. (2016). Necessary energy uses and a minimum standard of living in the United Kingdom: Energy justice or escalating expectations? *Energy Research & Social Science, 18*, 129–138.

CHAPTER 3

Practice and Energy Demand

Abstract This chapter introduces practice theory-inspired energy demand research focusing on key concepts and insights that speak to issues of inequality. The discussion explicates core ideas that have relevance in this respect, while also engaging with existing works that have sought to address questions of inequality from within practice-based energy research. The chapter then examines an important frontier in practice theory-inspired energy research of key relevance, I argue, for energy poverty—that of invisible energy policy. Finally, I move to draw together the different conceptual threads that have been laid out through the book thus far and raise key questions that emerge for analysis of energy poverty.

Keywords Practice theory · Energy demand · Inequality · Invisible energy policy

Introduction

The challenges of reducing energy demand and the development of approaches to understand energy use form the focus for a large body of work. Such work tends to be set within the broader context of environmental sustainability and particularly climate change. Indeed, central to the rationale for a research agenda on energy demand are the contentions that: (1) changes in forms of energy production (e.g. to renewable

C. Butler, *Energy Poverty, Practice, and Policy*, Progressive Energy Policy,
https://doi.org/10.1007/978-3-030-99432-7_3

energy technologies) are not sufficient to address associated environmental problems and (2) *reductions* in energy demand are required to meet existing commitments to reduce carbon emissions (Shove, 2015). These two basic arguments are broadly accepted by those working on energy demand issues, but beyond this there is vociferous debate. One of the key debates has a conceptual basis and concerns how social action can best be understood.

On the one hand, a fairly extensive body of research has sought to address questions of reducing energy demand with focus on psychological approaches to behavioural change (e.g. Dietz et al., 2009; Frederiks et al., 2015; Stern, 2000, 2020). On the other hand, a critically engaged analysis has come from the broadly termed 'practice tradition', with roots in sociological and geographical thought, where the basis for understanding social action implicit in much of the psychological literature is brought into question (e.g. Shove, 2003, 2010, 2011; Shove & Walker, 2014; Strengers & Maller, 2015). Crucially, where psychological behavioural research remains tied to dualisms of individual behaviour versus structural constraints, practice theory represents an attempt to characterise the interrelations between agency and structure. These fundamentally different conceptual approaches have been examined for their implications in understanding challenges associated with energy demand reduction, producing important insights.

Between these two poles of thought and analysis, there exists a spectrum of research that takes a weaker or stronger position on these conceptual issues (e.g. Gram-Hanssen, 2014; Spaargaren, 2011). Some propose alternatives, such as the energy cultures approach (e.g. Stephenson et al., 2015), while others have sought to engage with and address some of the key critiques levelled at psychological approaches (Whitmarsh et al., 2021). These offer important insights and developments for understanding behaviour change relevant to energy and its environmental consequences, as well as wider sustainability challenges. However, the focus for this book is on the practice tradition (e.g. Shove, 2003, 2010) with its conceptual emphasis on the complex relations that characterise the constitution of energy demand and needs for energy.

Here, I argue that there are core lines of thought within this tradition of energy demand research that offer important insights for energy poverty that as-yet have been left largely unexplored. Though practice theory-based analyses have offered deep understanding of the dynamics of energy demand, there has been little attention given to inequality or the implications of insights for energy poverty (Walker, 2013). This chapter discusses key ideas, contributions, and agendas from within practice-based energy research that I suggest have relevance for thinking about issues of energy poverty. These concern theorisation and analysis regarding how energy needs come into being and are actively constituted, and the emergent field of invisible energy policy as an important area of work that can speak to questions about the socio-political dynamics underpinning experiences of energy poverty. The chapter delineates these areas of practice-based energy demand research and moves towards a conclusion focused on drawing out the key questions that arise for energy poverty research and analysis.

Practice Theory in Energy Demand Research

The practice approach to energy demand takes a strong position on the importance of rejecting behavioural and cultural conceptualisations. Shove (along with several close collaborators) represents one of the foremost advocates of this approach. Centrally, Shove argues that 'instead of seeking more environmentally friendly ways of meeting given levels of service', through efficiency or behavioural interventions, more penetrating questions concern the processes through which services are specified and constituted in the first place (2003, p. 396). For Shove, the core question is: 'How do new conventions become normal, and with what consequence for sustainability?' (2003, p. 396). A later adjunct to this relates to the processes through which some practices are made obsolete or subject to decline, such as cycling (Shove et al., 2012). A by now well-known example, which Shove discusses in one of her earlier works, concerns practices of laundering.

In her 2003 paper, Shove offers an analysis that shows how contemporary conventions of laundering have co-evolved through the interaction of multiple mutually interdependent dimensions including technological development, conventions of cleanliness, and changes in clothing materials (notably the advent of synthetic materials). She highlights how the practice of laundering has at one time become less resource intensive, as processes of boiling have declined, but more demanding in that people are washing more frequently and combining washing with tumble drying. This kind of analysis, then, highlights how processes of change involve multiple interacting elements that lead to specifications of new forms of need with major implications (good or bad) for levels of energy demand.

This has been further developed in subsequent work to argue for a focus on 'what energy is for' within research and policy (Shove & Walker, 2014), rather than looking at energy as an abstract or underlying resource. As Shove and Walker (2014, p. 55) put it: 'energy is not used for its own sake but as part of accomplishing social practices'. Recognising this brings focus onto questions about how we live in ways that require energy use and how these particular configurations are constituted.

Shove is critical of existing focuses on efficiency and behavioural approaches because they 'obscure longer-term trends in demand and societal shifts in what energy is for' (2018, p. 779) and, as such, act to sustain increasingly energy-intensive ways of life. By way of an example these trends include things such as ongoing global increases in air conditioning. On this subject, Shove et al. (2013) offer a detailed analysis of the processes by which air conditioning is creating increasing demands for energy use, even in a context of broad recognition of needs to reduce energy demand. They argue that this is best explained by examining the ways that air conditioning has become embedded in specific forms of practice. In this regard, they highlight how the office environment has been fundamentally changed by a combination of computers, open plan spaces, and office wear, such that it increasingly involves air conditioning to cool equipment and space, as much as people. Or how the practices of nursing patients in intensive care have come to involve multiple new forms of technology such that air conditioning is now seen as a requirement of 'good' care. And how having air conditioning has become synonymous with quality in the hotel industry ramping up the requirements for these spaces to be routinely air conditioned (Shove et al., 2013). The analytic focus, then, is on the ways that practices come to require and depend upon ever higher needs for energy use.

This takes emphasis away from the individual as the unit of analysis towards practices themselves (Shove, 2010; Shove et al., 2012). Reflecting this position, Shove has developed a particular conception of practices that identifies three composite parts all of which require attention in processes of understanding how demands for energy come to be as they are. These three interrelated components have been identified as involving *materials*, *meanings*, and *competence* (Shove et al., 2012) or *material infrastructures*, *common understandings*, and *practical knowledge* (Shove & Pantzar, 2005). Though these elements have been characterised in slightly different ways, they offer a way into analysing and examining practices and the ways energy is implicated in what we do.

To give one example of how this can be applied analytically. Shove et al. (2012) use the practice of driving, working through the interrelationships between elements to show how these can be used to add to understanding of change. They point to *materials*, such as engines and carriages, *competences* such as mechanical expertise, steering, and braking, and *meanings*, such as exhibitions of wealth, links to adventure, work, and social relationships. This process of examining the interrelated elements that make up driving allows for examination of how they have changed over time. Crucially, by looking at the elements of practice, Shove et al. are able to show how the practice of driving had many precursors in daily life prior to the emergence of the car. They use this to highlight how the technology of the car and the materials that make up driving emerged and were moulded in interrelation with elements of competence and meanings.

These theoretical ideas about practice have been applied and developed further across a wide range of analyses (e.g. Hand et al., 2005; Hui et al., 2017; Maller & Strengers, 2013; Shove et al., 2012; Spurling & McMeekin, 2015; Strengers & Maller, 2015). These analyses have built insight into the ways that practices are formed and shaped within everyday life with implications for understanding the reduction of energy demand. For example, Maller and Strengers (2013) have shown how particular practices migrate with people as they move around the world, offering insights into the idea of practice memory and indicative of obduracy of practices across time and space, while Hui (2013) has developed a practice theory based analysis of mobilities giving particular focus to the ways that examining practice (in this case leisure practices) can be far more revealing for understanding the dynamics of travel, than examination of distances traversed, or time spent travelling. In focusing on practices, Hui makes a case for mobilities as inseparable from the leisure activities she examines (namely quilt making and bird watching). This brings a way of thinking about mobilities as embroiled in multiple practices and the elements that compose them, rather than looking at them as something distinctive by focusing on flying or driving, for example.

Other contributions have sought to engage with practice theory through a focus on the person, as opposed to the practice. This has been central to debates about method and how we can empirically research practices, taking a position that some formulations of conventional social scientific methods, such as interviews, can be utilised in ways consistent with practice theory (Butler et al., 2014; Hitchings, 2011). A key focus for this research has been on life narratives and biographical trajectories and the insights scrutiny of these can provide into how practices are shaped through time (e.g. Butler et al., 2014; Fox et al., 2017; Groves et al., 2016; Hards, 2012). For example, in my own work (Butler et al., 2014, 2016), I have examined how people's biographical experiences through their life courses intertwine with and reproduce social and material structures of consequence for energy demand. Here, the focus has similarly been on the interrelationships between different elements of change but with an approach that explores the ways agency-structure interrelations are constituted over time within and through people's lives.

The wide-ranging body of work discussed thus far offers some key concepts and inroads for thinking about inequality and poverty. However, they have rarely been applied in this way and very little research has developed practice-based thinking with focus on energy poverty. I argue that there is, however, important insight to be gained from using practice-inspired analysis in understanding energy deprivation. Within the literature, there are a small number of works and references to issues that span practice and inequality, with a few making further connections through to energy poverty. It is to discussion of these that I now turn.

Bringing Inequality into Practice Theory: Key Concepts and Interventions

> It is… hard to find examples of research that is inspired by theories of social practice and that explicitly addresses the reproduction of abject poverty, that analyses the failure to successfully perform everyday practices, or that directly engages with the reproduction of social inequality and justice. (Walker, 2013, p. 181)

This quote from Walker highlights how despite the importance of practice theory-inspired analyses for energy demand research, the relevance of inequality has been largely neglected. Here, I focus on the small number of existing interventions that have sought to bring practice theory to bear on questions of inequality, suggesting connections across to issues of energy poverty. Of particular interest for present purposes are concepts of *recruitment*, *defection*, and *reproduction*, along with notable interventions from Shove (2002) and Walker (2013).

Recruitment and defection refer to the ways that people—defined as carriers of practice—can be recruited to or defect from practices, effectively shaping which practices are reproduced and sustained through time and which decline (see Shove et al., 2012). One key example developed by Shove et al. (2012) is that of showering as a practice that has successfully recruited large populations of people into enacting it every day. They use examples like this to set out the dimensions of practice which appear to successfully recruit practitioners, including the embedding of practices with infrastructures, institutions, and norms, but also things like the possibilities for innovation and the 'rewards' (both internal and external) that different practices afford. Central to their argument here is the idea that 'people are unknowingly engaged in reproducing and enacting multiple and varied cycles of change, simultaneously shaping the lives of practices and being shaped by them' (Shove et al., 2012, p. 77).

Crucially, it is in the elaboration of these concepts that the issue of inequality is introduced, though not explored in detail nor developed in relation to ideas of energy poverty. Shove et al. (2012, p. 65) recognise that 'social and material inequalities restrict the potential for one or another practice to develop' and limit the chances that people have for becoming carriers of any one practice. However, they focus their analysis on the ways that practices are developed and sustained or decline and expire owing to the extent to which cohorts of practitioners enact them, rather than the possibilities of exclusions from practices that arise due to inequalities. Additionally, questions about differences between practices in terms of the extent to which practitioners are compelled to engage in them or not, are hinted at in a brief mention of law but left broadly unaddressed.

Walker (2013)—as highlighted in the quote at the opening to this section—offers a more detailed examination of these questions, taking forward a conceptual analysis that draws social practice theory [using Schatzki's work] into dialogue with a specific strand of justice thinking [Sen's capabilities approach]. The key issues that he brings to light through his analysis concern, first, the differential capabilities that people have for performing different practices successfully. Or drawing on Sen, the different capabilities and potential they have for enacting social practices (the capabilities approach to energy poverty is discussed in Chapter 2). And second, the ways in which patterns of 'recruitment' to and 'defection' from practices can be contentious.

He asserts that often discussions of recruitment to and defection from practices can make such processes appear unproblematic, referring often to leisure pursuits (Hui, 2013; Shove et al., 2012). But if the practice in question is one to which access is restricted in some way, they can appear far more normatively charged. Walker (2013) points out that many practices-as-entities have embedded rules and norms and they make certain physical or material demands that restrict opportunities to participate. In this way, recruitment can appear as inseparable from capability, as if a particular practitioner lacks the capabilities to fulfil a practice, they are unrecruitable and effectively excluded.

Equally, Walker (2013) suggests that a similar line of argument can be applied to defection, with people having varying degrees of choice over whether and how they defect from practices according to their capabilities. He offers the example of a person that defects from driving because of deteriorating health or loss of employment to illustrate. This analysis offers particular focus, then, on the ways that issues of inequality can be central to questions of recruitment and defection from practices, raising valuable conceptual openings that are pertinent for thinking across practice and poverty in the energy context.

This has been touched on elsewhere in work using practice perspectives, where the relevance of looking at variation in the experiences of different groups offers a further line of thinking for engaging with inequality. Fox et al. (2017) take forward a practice-based analysis of people in later life examining how life experiences within an older demographic are shaping energy demand trends in travel (i.e. towards increasing demand). They emphasise 'how travel desires come about through the production of certain shared expectations, aspirations and other normative dimensions' (Fox et al., 2017, p. 105). But they also draw out the importance of *personal and corporeal capacity* within consideration of the recruitment of people to practices, highlighting how there are differences in capabilities to carry out practices. In this respect, their focus is on ageing bodies, but this insight brings further questions about the role of inequality in processes of recruitment and defection.

Shove has also—in earlier work—more deeply grappled with questions of inequality using practice theory to theorise social exclusions relating to mobilities (Cass et al., 2005; Shove, 2002). In this conceptualisation, social exclusion is positioned as an emergent property of three elements; (1) social practices and the obligations to perform them; (2) individual resources and capacities to meet obligations; and (3) infrastructures that shape people's abilities to meet obligations as well as the expectations of 'normal' social participation. This brings in possibilities for thinking about how inequality is woven through these different intersecting elements. For instance, abilities to meet obligations are related to the extent of available resources and the level of access to infrastructures, both of which are unequally distributed across societies. But it also directs attention to the ways that obligations and abilities are created. Shove highlights the role of policy, in particular, as being 'deeply implicated in the construction of...demand and in the shaping of social expectations and practices' (2002, p. 10). This hints at the importance of power in the construction of practices and the requirements for participation—something that I argue could form a far more explicit and important area for analysis in energy poverty research and, as such, is given attention in the empirical analysis in the later parts of this book.

Focusing as it does on mobilities, this work (and others) also speaks to the ways that practice theory thinking can reorient analysis of transport poverty. Centrally, it shifts focus away from *enabling access* to transport, towards questions about why people travel (i.e. to enact which practices), and the ways that particular mobilities are constructed (e.g. through the favouring of infrastructure for car travel) (Cass et al., 2005; Hui, 2013; Mullen & Marsden, 2016; Shove, 2002). The notion of travel being about getting from A to B is supplanted by thinking of it as intricately woven into the accomplishment of practices. As Shove puts it, mobility... is about integrating everyday life and the activities required of 'normal practice' (2002, p. 9). This calls attention to what comes to be regarded as 'normal', and to how and why practices, and the requirements for mobility that they entail, come to be as they are. With this at the fore, transport poverty can be thought about in very different terms, with less focus on enabling access and more thought given to the constitution of needs.

Overall, the contributions, debates, and analyses from practice theory-based energy research have had major implications for the ways that energy demand is thought about and addressed. They have moved focus away from individualised decisions and choices towards the interrelationships between human agency and socio-material structures in shaping processes of change. They have highlighted the need to think beyond technical efficiency and economic rationality to bring into focus processes that are contributing to global increases in energy demand. And, crucially, they have brought attention to questions concerning 'what energy is for', highlighting the ways in which demand for energy has been, and continues to be, constituted by processes of governance. Though there are some key interventions and nods to issues of inequality within this literature, I argue this remains an area that could be advanced much further and with greater attention across to energy poverty.

This chapter moves towards extrapolating key questions that arise for thinking about energy poverty when practice theory ideas are brought into play. Thus far, key concepts and insights related to the constitution of need have been foregrounded, but the next section delineates the other area of practice theory-based analysis that I suggest has potential for exploring challenges of energy poverty—that of invisible energy policy. Here, the relevance of governance and policies in shaping and shifting practices across diverse areas of policy far beyond energy is made the focus. This burgeoning literature has thus far primarily sought to emphasise and trace the connections between policy, practices, and environmental sustainability, but I argue offers an equally important agenda for energy poverty research.

Practice Theory and the Emergence of Invisible Energy Policy

The fundamental contention of the invisible energy policy literature is that when focus is brought onto practices and *what energy is for*, it becomes possible to see how multiple areas of policy far beyond energy have implications for shaping, shifting, and instituting demand (e.g. Butler et al., 2018; Cox et al., 2019; Royston et al., 2018). Although the role of policy in constituting needs for energy has been identified within several practice-inspired studies of energy demand, this had not previously formed an explicit focus of analysis. The wider influence on energy demand of policies from areas as diverse as health, work, education, and housing is identifiable but very rarely is this at the fore. This emerging body of work seeks to address this gap by bringing focus on key areas of policy that, while identifiable as having important implications for energy demand, have yet to receive sufficient attention (Cox et al., 2019).

The work in this space has set out some key tenets around which a research agenda has begun to be formulated. First, the idea that policies can be regarded as invisible where they are explicitly designed to address policy priorities outside of energy but nonetheless have impacts on energy demand and issues. 'Invisibility' in this context is referring to the ways that such impacts are either 'unacknowledged or insufficiently acknowledged' (Royston et al., 2018, p. 128), somewhat softening the intended implications of the 'invisible' concept. And the boundaries between what constitutes visible or invisible energy policy are also recognised as being 'complicated and blurred' (Royston et al., 2018, p. 128). Second, the links between policies and their impacts for practice and energy demand are recognised as being non-linear, complex, and varying from direct to indirect in the forms their effects take (Butler et al., 2018).

Of course, one might expect to find resources for these more cross-cutting analyses of governance within existing literatures and disciplines, such as political science. But Royston et al. (2018) have argued convincingly that these do not necessarily offer a good grounding for examining the kinds of issues with which this agenda is concerned. For example, while political scientists are adept at addressing policy causes and effects more broadly, they have given relatively little attention to specific functional policy areas, meaning that policy processes relating to an area like energy demand have rarely been addressed (Royston et al., 2018; and for an exception see Kuzemko et al., 2017). Equally, when the focus is on particular policies, as in impact assessments, the remit is often extremely narrow excluding possibilities for looking across the broader policy spectrum or for thinking about long-term trends. And where analysis is directed at issues of joined-up policy (Davies, 2009), studies tend to focus on the governmental processes involved, rather than attempting to understand the lived experiences of policy and their implications for practice.

At the same time, the fields of political economy and political ecology frequently address processes of global and local energy and environmental crises and offer explanatory power in terms of the role of political forces in shaping them, but tend to be either rooted in structuralist conceptual traditions that focus on political and economic power and the regulatory processes within governments (e.g. Mitchell, 2008), or engaged in post-structural analysis of altered subjectivities in the face of global economic processes, but again rarely tuned into the specifics of policies (e.g. Escobar, 1996). It is possible to assert, therefore, that there are limited available resources or methods within established traditions for investigating precisely how non-energy policies shape practice with consequences for energy demand issues.

Royston et al. (2018) make a case for a more ambitious agenda that seeks to address; 'more fundamental questions about the changing array of 'services' that energy makes possible, about the amount of energy 'needed' in society, [and] about the role of policy in constituting these 'needs'' (Royston et al., 2018, p. 127). While, as noted, some conceptual traditions might situate governance as some form of driver or external influence on social action, understandings consistent with practice theory bring a different orientation—one which is more attentive to complexity and non-linearity (e.g. see Butler et al., 2018; Urry, 2010), while also recognising processes by which practices are shaped and shifted.

Arising from these wider conceptual developments and agenda-setting papers are a number of studies of invisible energy policy focused on different policy areas, including health (Blue, 2017; Nicholls & Strengers, 2018), digitalisation (Morley et al., 2018), and education (Gormally et al., 2019; Royston, 2016), as well as my own work on welfare policy (Butler et al., 2018). There are also studies looking at the ways that multiple different policy areas intersect within daily life to shape domestic energy practices, moving outside of the focus on specific policy areas that have dominated elsewhere (Greene & Fahy, 2020). As a body of research, this has highlighted the ways that invisible energy policies are constitutive of new needs for energy demanding services (e.g. Butler et al., 2018; Morley et al., 2018; Nichols & Strengers, 2018); how they can have direct impacts on energy issues and practices, as well as much longer-term and indirect forms of impact, for instance in shaping how energy issues are framed or delimited (e.g. Butler et al., 2018); how boundaries within governance processes can be constitutive of in/visibility (e.g. Cox et al., 2019); and the ways in which the demands of different policy agendas are negotiated by people in and through practice (Gormally et al., 2019).

Though the agenda-setting papers in the invisible energy policy space include scope for examining energy poverty and to some extent highlight it as an area for analysis (Cox et al., 2019), it is fair to say that the focus to date has primarily been on issues of energy demand reduction (with some notable exceptions, e.g. see Butler et al., 2018; Nicholls & Strengers, 2018). This is likely because the concern with invisible energy policy has emerged from the practice theory-inspired literature that as highlighted has focused on energy demand reduction and sustainability with far less consideration of inequality and issues of energy poverty.

Equally, the differing concepts found in much of the energy poverty literature (e.g. vulnerabilities, capabilities, precarity) have tended to mean invisible energy policy has not been taken up as readily by scholars working on these issues. Though there are a small number of studies that look at the impact of welfare reform on fuel poverty (e.g. Snell et al., 2015), these tend not to engage with the wider practice theory literature in which the invisible energy policy agenda is grounded. This means that though they identify areas of impact, they do not bring focus on the conceptual insights afforded by working with practice theory and energy poverty together.

Across these key areas of practice theory-inspired research, I argue there are important ideas that can be brought to bear in analysis and thought about energy poverty. The existing interventions that foreground questions of inequality and practice signal possibilities for thinking about the constitution of needs in the context of energy poverty. And the invisible energy policy literature brings closer attention to the role of policies far beyond energy policy in both processes of constitution and calling attention to how such policy is shaping experiences of energy poverty. The conclusion of this chapter serves to draw these conceptual insights from practice theory together with those advanced in the previous chapter around energy poverty, capabilities, and precarity, to introduce key lines of enquiry for this research area.

Conclusion

This chapter has discussed the practice theory-inspired literature on energy demand drawing together key ideas and concepts that arise from this tradition for thinking about energy poverty and inequality. Two major areas of thought have been examined first, ideas about *the constitution and specification of need* involving processes of recruitment to and defection from practices, and second, the *invisible energy policy agenda* where the effects of policy outside of energy are given focus. I argue that these different areas of conceptual development can be brought into closer conversation with concepts from energy poverty research to inform analytic endeavours and research agendas.

First, the capabilities approach to energy poverty has already been cast in terms of how it affords possibilities for exploration of the ways that societal norms shape energy service needs. At present, thinking in terms of energy poverty more widely often focuses on the ways that energy service needs can be met in some way. Day et al. (2016), however, denote an understanding of energy poverty derived from capabilities that begins to call in to question the specification of needs for energy in a similar way to that found in practice theory-inspired energy research. They bring focus on how social norms shape the relations between energy services and capabilities with implications for understanding energy deprivation. For example, they highlight how showering every day as a normal expectation for people in the UK can be cast in terms of its importance to maintaining the capability of social respect. This marks out showering, then, as an energy service that should be considered within energy poverty analysis in the UK context because of its links to basic capabilities. Such an approach thus offers a way of thinking about the links between social norms, energy services, and capabilities but it also signals potential to go further and bring attention to how different energy service needs come to be *made* essential to capabilities, i.e. by asking—in line with practice theory-inspired work—how social norms come to be as they are.

In this respect, I suggest a capabilities approach is well suited to alignment with the concerns of practice theory-inspired energy analysis. While analysis of capabilities brings focus on the implications of energy service needs for energy poverty, practice theory concepts force attention onto how those needs are created and, moreover, invisible energy policy insights emphasise the role of diverse policy areas in such processes of constitution. Within this, key concepts of *recruitment* and *defection* from practice theory can be used to frame questions about inequality in the processes through which people become enrolled in energy demanding practices and related social norms. Such questions concern who has the power to constitute needs and how do abilities to resist, be recruited, or defect from new norms of practice vary across different people and policy areas. All this speaks to openings for an analysis of energy poverty that places the increasing energy intensity of daily life more firmly at the heart of debates.

Second, beyond offering understanding of the constitution of needs, the invisible energy policy agenda has further value for extending thinking about how experiences of energy poverty are being shaped by non-energy policy areas. This line of analysis is more concerned with examining the ways that non-energy policy affects the prevailing conditions for energy poverty, with less focus on the constitution of need and more attention to other important points of intersection. For example, work from within the energy poverty literature building from concepts of vulnerability and precarity has already signalled the importance of social and political processes emanating from non-energy policy areas, such as housing and welfare, in shaping experiences of energy deprivation (Middlemiss, 2016; Petrova, 2018). This research highlights complex forms of influence that shape things like the normalisation of poor housing for some demographics or what can be counted as fuel poverty as distinct from wider poverty. With invisible energy policy as a starting point for analysis, these types of concerns can be foregrounded with potential to advance understanding of energy poverty in ways that extend beyond the preoccupations and concerns of fuel poverty policy.

In the following two chapters, the areas of governance and the policy contexts within which the book's analysis is situated are discussed. Though the empirical research in this book is concerned with welfare policy as an area of invisible energy policy, it is nonetheless important to discuss wider energy demand policy too. This is revealing for understanding how definitions of energy demand issues, across poverty and demand reduction, are characterised in UK policy. Chapter 4 thus examines the UK energy demand and fuel poverty policy context before moving to address welfare policy and discuss existing connections across these policy areas. In the remaining Chapters (5 and 6), attention turns to the empirical research and its exploration in relation to the areas of theory and conceptual contribution discussed here. The ideas advanced through discussion of the literature across energy poverty and practice theory-inspired energy research are developed further, and the empirical materials are used to explore avenues of analysis at the intersections.

References

Blue, S. (2017). Reducing demand for energy in hospitals: Opportunities for and limits to temporal coordination. In A. Hui, R. Day, & G. Walker (Eds.), *Demanding energy: Space time and change* (pp. 313–338). Palgrave.

Butler, C., Parkhill, K. A., & Luzecka, P. (2018). Rethinking energy demand governance: Exploring impact beyond "energy" policy. *Energy Research & Social Science, 36*, 70–78.

Butler, C., Parkhill, K. A., & Pidgeon, N. F. (2016). Energy consumption and everyday life: Choice, values and agency through a practice theoretical lens. *Journal of Consumer Culture, 16*(3), 887–907.

Butler, C., Parkhill, K. A., Shirani, F., Henwood, K., & Pidgeon, N. (2014). Exploring the dynamics of energy demand through a biographical lens. *Nature and Culture, 9*(2), 164–182.

Cass, N., Shove, E., & Urry, J. (2005). Social exclusion, mobility and access. *The Sociological Review, 53*(3), 539–555.

Cox, E., Royston, S., & Selby, J. (2019). From exports to exercise: How non-energy policies affect energy systems. *Energy Research & Social Science, 55*, 179–188.

Davies, J. S. (2009). The limits of joined-up government: Towards a political analysis. *Public Administration, 87*(1), 80–96.

Day, R., Walker, G., & Simcock, N. (2016). Conceptualising energy use and energy poverty using a capabilities framework. *Energy Policy, 93*, 255–264.

Dietz, T., Gardner, G. T., Gilligan, J., Stern, P. C., & Vandenbergh, M. P. (2009). Household actions can provide a behavioral wedge to rapidly reduce US carbon emissions. *Proceedings of the National Academy of Sciences, 106*(44), 18452–18456.

Escobar, A. (1996). Construction nature: Elements for a post-structural political ecology. *Futures, 28*(4), 325–343.

Fox, E., Hitchings, R., Day, R., & Venn, S. (2017). Demanding distances in later life leisure travel. *Geoforum, 82*, 102–111.

Frederiks, E. R., Stenner, K., & Hobman, E. V. (2015). Household energy use: Applying behavioural economics to understand consumer decision-making and behavior. *Renewable and Sustainable Energy Reviews, 41*, 1385–1394.

Gormally, A. M., O'Neill, K., Hazas, M. D., Bates, O. E. G., & Friday, A. J. (2019). "Doing good science": The impact of invisible energy policies on laboratory energy demand in higher education. *Energy Research & Social Science, 52*, 123–131.

Gram-Hanssen, K. (2014). New needs for better understanding of household's energy consumption—Behaviour, lifestyle or practices? *Architectural Engineering and Design Management, 10*(1–2), 91–107.

Greene, M., & Fahy, F. (2020). Steering demand? Exploring the intersection of policy, practice and lives in energy systems change in Ireland. *Energy Research & Social Science, 61*, 101331.

Groves, C., Henwood, K., Shirani, F., Butler, C., Parkhill, K., & Pidgeon, N. (2016). Energy biographies: Narrative genres, lifecourse transitions, and practice change. *Science, Technology, & Human Values, 41*(3), 483–508.

Hand, M., Shove, E., & Southerton, D. (2005). *Explaining showering: A discussion of the material, conventional, and temporal dimensions of practice* [online] www.socresonline.org.uk, http://www.socresonline.org.uk/10/2/hand.html. Accessed 2 September 2021.

Hards, S. (2012). Tales of transformation: The potential of a narrative approach to pro-environmental practices. *Geoforum, 43*(4), 760–771.

Hitchings, R. (2011). People can talk about their practices. *Area, 44*(1), 61–67.

Hui, A. (2013). Moving with practices: The discontinuous, rhythmic and material mobilities of leisure. *Social & Cultural Geography, 14*(8), 888–908.

Hui, A., Schatzki, T., & Shove, E. (Eds.). (2017). *The Nexus of practices: Connections, constellations, practitioners*. Routledge.

Kuzemko, C., Mitchell, C., Lockwood, M., & Hoggett, R. (2017). Policies, politics and demand side innovations: The untold story of Germany's energy transition. *Energy Research & Social Science* [online], *28*, 58–67.

Maller, C., & Strengers, Y. (2013). The global migration of everyday life: Investigating the practice memories of Australian migrants. *Geoforum, 44*, 243–252.

Middlemiss, L. (2016). A critical analysis of the new politics of fuel poverty in England. *Critical Social Policy, 37*(3), 425–443.

Middlemiss, L., Ambrosio-Albalá, P., Emmel, N., Gillard, R., Gilbertson, J., Hargreaves, T., Mullen, C., Ryan, T., Snell, C., & Tod, A. (2019). Energy poverty and social relations: A capabilities approach. *Energy Research & Social Science, 55*, 227–235.

Mitchell, C. (2008). *The political economy of sustainable energy*. Palgrave Macmillan.

Morley, J., Widdicks, K., & Hazas, M. (2018). Digitalisation, energy and data demand: The impact of Internet traffic on overall and peak electricity consumption. *Energy Research & Social Science, 38*, 128–137.

Mullen, C., & Marsden, G. (2016). Mobility justice in low carbon energy transitions. *Energy Research & Social Science, 18*, 109–117.

Nicholls, L., & Strengers, Y. (2018). Heatwaves, cooling and young children at home: Integrating energy and health objectives. *Energy Research & Social Science, 39*, 1–9.

Petrova, S. (2018). Encountering energy precarity: Geographies of fuel poverty among young adults in the UK. *Transactions of the Institute of British Geographers, 43*(1), 17–30.

Reckwitz, A. (2002). Toward a theory of social practices. *European Journal of Social Theory, 5*(2), 243–263

Royston, S. (2016). Invisible energy policy in Higher Education. In *Proceedings of the Demand Conference* (pp. 13–15).

Royston, S., Selby, J., & Shove, E. (2018). Invisible energy policies: A new agenda for energy demand reduction. *Energy Policy, 123*, 127–135.

Schatzki, T. (1996). *Social practices: A Wittgensteinian approach to human activity and the social*. Cambridge University Press.

Shove, E. (2002, October). *Rushing around: Coordination, mobility and inequality*, draft paper for the Mobile Network meeting.

Shove, E. (2003). Converging conventions of comfort, cleanliness and convenience. *Journal of Consumer Policy, 26*(4), 395–418.

Shove, E. (2010). Beyond the ABC: Climate change policy and theories of social change. *Environment and Planning A: Economy and Space, 42*(6), 1273–1285.

Shove, E. (2011). On the difference between Chalk and Cheese—A response to Whitmarsh et al's comments on "beyond the ABC: Climate change policy and theories of social change." *Environment and Planning A: Economy and Space, 43*(2), 262–264.

Shove, E. (2015). Linking low Carbon policy and social practice. In Y. Strengers & C. Maller (Eds.), *Social practices, intervention and sustainability: Beyond behaviour change* (pp. 31–44). Routledge.

Shove, E., & Pantzar, M. (2005). Consumers, producers and practices. *Journal of Consumer Culture, 5*(1), 43–64.

Shove, E., Pantzar, M., & Watson, M. (2012). *The dynamics of social practice: Everyday life and how it changes*. Sage.

Shove, E., & Walker, G. (2014). What Is energy for? Social practice and energy demand. *Theory, Culture & Society, 31*(5), 41–58.

Shove, E., Walker, G., & Brown, S. (2013). Transnational transitions: The diffusion and integration of mechanical cooling. *Urban Studies, 51*(7), 1506–1519.

Snell, C., Bevan, M., & Thomson, H. (2015). Welfare reform, disabled people and fuel poverty. *Journal of Poverty and Social Justice, 23*(3), 229–244.

Spaargaren, G. (2011). Theories of practices: Agency, technology, and culture. *Global Environmental Change, 21*(3), 813–822.

Spurling, N., & McMeekin, A. (2015). Interventions in practices: Sustainable mobility and policies in England. In Y. Strengers & C. Maller (Eds.), *Social practices, intervention and sustainability: Beyond behaviour change* (pp. 78–94). Routledge.

Stephenson, J., Barton, B., Carrington, G., Doering, A., Ford, R., Hopkins, D., Lawson, R., McCarthy, A., Rees, D., Scott, M., Thorsnes, P., Walton, S., Williams, J., & Wooliscroft, B. (2015). The energy cultures framework: Exploring the role of norms, practices and material culture in shaping energy behaviour in New Zealand. *Energy Research & Social Science, 7*, 117–123.

Stern, P. C. (2000). New environmental theories: Toward a coherent theory of environmentally significant behavior. *Journal of Social Issues, 56*(3), 407–424.

Stern, P. C. (2020). A reexamination on how behavioral interventions can promote household action to limit climate change. *Nature Communications, 11*(1).

Strengers, Y., & Maller, C. (Eds.). (2015). *Social practices, intervention and sustainability: Beyond behaviour change*. Routledge.

Urry, J. (2010). Mobile sociology. *The British Journal of Sociology, 61*, 347–366.

Walker, G. (2013). Inequality, sustainability, and capability: Locating justice in social practice. In E. Shove & N. Spurling (Eds.), *Sustainable practices: Social theory and climate change* (pp. 181–196). Routledge.

Watson, M. (2017). Placing power in practice theory. In A. Hui, T. Schatzki, & E. Shove (Eds.), *The Nexus of practices: Connections, constellations and practitioners* (pp. 169–182). Routledge.

Whitmarsh, L., Poortinga, W., & Capstick, S. (2021). Behaviour change to address climate change. *Current Opinion in Psychology, 42*(76), 81.

CHAPTER 4

Policy: Energy Demand and Welfare in the UK

Abstract This chapter presents a detailed discussion of contemporary UK energy demand policy and welfare policy. These two areas of policy form the focus of the empirical research that will be utilised to examine and develop the conceptual ideas discussed in Chapters 2 and 3. Through this chapter, I first discuss policies relating to fuel poverty, contextualising them in relation to wider energy demand strategies, before going on to address welfare policy with focus on contemporary reforms. In concluding the chapter, I explore existing forms of interconnection for these different policy areas and issues, explicating some of the key challenges that arise for thinking across different domains.

Keywords Invisible energy policy · Energy demand policy · Fuel poverty policy · Welfare policy

Introduction

Fuel poverty, as a distinctive policy problem, is a relatively recent phenomenon tied to a particular set of social and political projects, as well as specific metrics and social groups that were initially the focus in bringing it onto the public agenda. Many countries around the world still do not recognise fuel or the broader category of energy poverty as a specific issue (Bouzarovski & Petrova, 2015). Correspondingly, it does

C. Butler, *Energy Poverty, Practice, and Policy*, Progressive Energy Policy,
https://doi.org/10.1007/978-3-030-99432-7_4

not form a focus for analysis, funding, or associated policy initiatives. The UK was one of the first nations to recognise fuel poverty as a distinctive policy problem and dedicate resources and policy attention to addressing the associated challenges (Bouzarovski, 2018). This has meant it has been particularly influential in the development of fuel poverty policy around the world (Mattioli et al., 2017), making a focus on the UK interesting beyond the specific country context (Bouzarovski, 2018). Though the focus within this book and chapter is on fuel poverty policy, it is necessary to contextualise this area in relation to wider energy demand strategy since the two are intimately interlinked.

Over the past several years, there have been some important developments in energy policy agendas related to energy demand. The UK's Climate Change Act (2008), which enshrined in law a target to reduce greenhouse gas emissions by 80% (relative to 1990 levels) by 2050, has been amended to implement a new 'net zero' target (Parliament UK, 2019). At the same time, the current UK Government has been criticised for 'dismantling' much of the policy intended to ensure delivery of such targets (Parliament UK, 2019). Indeed, the Climate Change Committee (2019, p. 11) has stated that 'current policy is insufficient for even the existing targets'. Crucially, these targets are asserted to require a far greater focus on energy demand if they are to be met (Eyre & Killip, 2019).

Since 2010, however, shifting political agendas have seen fuel poverty policy being prioritised over energy demand reduction, with alterations made to the primary energy demand policy—the Energy Company Obligation scheme (ECO). This scheme is characterised by a market-led approach that focuses primarily on improving housing efficiency and, while originally it was aimed at all housing, it has increasingly shifted to be focused on the worst housing and those on low incomes. The nature of UK fuel poverty policy is thus directly tied to a strategy initially designed to address energy efficiency across all forms of housing.

With the UK's climate leadership role as hosts of the Conference of Parties (COP) 2021 a newly renewed focus on climate policy had begun to emerge, but there have also been further moves towards fuel poverty being the principle focus for energy demand policy with the publication of the *Sustainable Warmth* strategy (BEIS, 2021). Though the shift to focus policy on fuel poverty is important and necessary, little has arisen to address energy demand reduction issues more broadly. These shifts are important to highlight to provide context for the nature of fuel poverty policy in the UK. The chapter first offers an examination of these moves toward an increasing emphasis on fuel poverty within energy demand policy.

The shifting landscape of energy demand policy also gives insight into definitions and understandings of fuel poverty, with a relatively narrow focus on housing efficiency and heat evident in official narratives. In UK policy, this is reflected in changes in the specific approach to defining fuel poverty and the targeting of measures to enhance efficiency (BEIS, 2021; Hills, 2012). In this chapter, I discuss the dominant policy definitions and highlight how such approaches are in distinct contrast with the wider framings of energy poverty delineated in the academic literature and discussed in Chapter 2 (e.g. Bouzarovski & Petrova, 2015; Day et al., 2016; Middlemiss & Gillard, 2015).

The chapter finally moves to address welfare policy as the area of 'invisible energy policy' (i.e. non-energy policy areas that have implications for energy demand issues) that forms the focus of the research. I examine existing links, connections, and boundaries with fuel poverty policies, and set out the policies, framings, and politics that dominate in this area. The chapter highlights how there are interconnections across to welfare policy but these are narrowly conceived and offer relatively little scope for more reflective thinking across different areas. While the invisible energy policy literature (e.g. Butler et al., 2018; Cox et al., 2019; Greene & Fahy, 2020; Royston et al., 2018) directs analysis towards links and interconnections, there are clear limitations on the extent to which these are (or even can be) currently considered within existing policy landscapes.

From Energy Demand Reduction to Fuel Poverty Policy

Energy demand reduction has been highlighted as central to delivering the net zero carbon emissions target and for meeting sustainability goals more widely. In a report on UK energy demand policy, Eyre and Killip (2019) assert that 'Energy demand reduction, flexibility and decarbonisation will need to play a critical role [in sustainable transitions] and this should be recognised in energy innovation policy' (2019, p. 7). At the same time, they point to inconsistency in policy development and implementation, with many successful policies having been reduced in scale or abandoned in recent years (Eyre & Killip, 2019). These include the Carbon Emissions Reduction Target (CERT), the

Community Energy Savings Programme (CESP), the Carbon Reduction Commitment (CRC), and the proposed Zero Carbon Homes (ZCH) standard. These latter two policies were aimed at large organisations and new build housing, while the former two were targeted at the domestic sector. While the CRC and ZCH policies were ended with the election of a new conservative/liberal coalition government in 2010, the CERT and CESP policies were replaced with a similar but distinct mechanism, namely the Energy Company Obligation (ECO).

CERT required larger gas and electricity suppliers to achieve targets for reducing carbon emissions from domestic premises. The policy, which was enshrined in the Electricity and Gas (Carbon Emissions Reductions) Order (2008) and subsequent amendments, specified both the level of savings required and the ways that these were to be achieved—in particular, and increasingly as the scheme went on, the focus was on insulating homes. The target was set at 293 million lifetime tonnes of carbon dioxide emissions by December 2012. Suppliers delivered savings of 296.9 Mt CO_2 by the end of the scheme, fulfilling the policy aims and causing it to be recognised as a success.

CERT also gave focus to a 'priority group' and required that at least 40% of the target had to be achieved through uptake by people and households within this group. There were further additional requirements to target a 'super priority group', which was defined as including people that qualified for welfare benefits, such as child tax credits, and below an income threshold. These requirements embedded an aim to align carbon reduction measures with addressing energy vulnerability and fuel poverty, though some argue it did not go far enough with the majority of measures still being deliverable to households outside of the priority groups (CSE, 2014).

CESP was also an obligation on suppliers to deliver energy saving measures but targeted specifically at low-income areas. This scheme was designed to promote a 'whole house' approach and to treat as many properties as possible within geographically defined areas selected using measures of multiple deprivation. Under this scheme, energy companies achieved over 85% of the target savings. Despite having been regarded as successes, in 2012 CERT and CESP were replaced by the most recent policy addressing energy demand—the Energy Company Obligation.

This policy continued in a similar vein to CERT and CESP by placing obligations on energy suppliers to deliver carbon emissions reductions or energy savings but with some marked differences. For example, under

ECO, greater emphasis is placed on the fuel poor and 'hard to reach' properties, and companies have a larger degree of freedom in *how* they deliver on their obligations (i.e. which measures they implement). The components of ECO at its inception were as follows: the *carbon emissions obligation* is a requirement for energy companies to promote free instalment of energy saving measures (this includes loft and cavity wall insulation but there is no requirement as under CERT to prioritise this measure). The *carbon saving community obligation* targets low-income areas and promotes insulation more directly along with other measures such as connections to district heating systems. Finally, the *home heating cost reduction obligation* is an initiative targeted at improving the ability of low-income and vulnerable households to effectively heat homes. This typically includes measures such as boiler replacement.

ECO has since been amended through three further phases ECO2 (April 2015–March 2017) and an extension ECO2t (April 2017–September 2018), ECO3 (October 2018–March 2022) and ECO4 (April 2022–March 2026). Over this time, the changes to ECO have seen a move toward focusing on fuel poor households, with ECO3 being dedicated in its entirety to support for low income, vulnerable and fuel poor households and effectively ending the *carbon emissions obligation* component. This move has been regarded as progressive by many of those working in fuel poverty. However, it does leave a gap in terms of energy demand reduction policy for those outside of low-income groups. This, of course, does not require redirection of funding from ECO away from fuel poverty, but rather distinct and new policy initiatives that are able to address the major challenges facing the UK in reducing energy demand.

ECO was introduced initially in conjunction with the Green Deal—a financing mechanism designed to deliver energy efficiency retrofit to housing. Households could apply for loans toward the cost of home energy-efficiency measures, which would be repaid through the savings made on energy bills owing to the improvements. The policy was deemed unsuccessful due to low take-up and concerns about the quality of installations and was abandoned in 2015. With nothing to replace the Green Deal, and with the end to wider policies such as Zero Carbon Homes, this has left a lacuna in energy demand reduction policy.

Several other policy proposals and initiatives have failed to get off the ground or have been very short lived. One of the most prominent being the Green Homes Grant scheme that saw grants of between £5000 and

£10,000 offered to fund up to two-thirds of the cost of energy efficiency and sustainability improvements (e.g. insulation, double glazing, heat pumps and biomass boilers) but was abolished after only 6 months amid major problems with the administrative complexity of the scheme for householders and contractors alike. The only lasting policy in this area since 2010 is the minimum energy efficiency standard for the Private Rented Sector (BEIS, 2019, 2021). This sets a minimum energy efficiency standard for landlords wherein all privately rented properties must receive an Energy Performance Certificate (EPC) of band E or above. The aim is to tackle the least energy-efficient (bands F or G) private rental properties in England and Wales. These minimum standards have also become an increasing focus within fuel poverty policy amid recognition that '33.6% of fuel poor households in England are living in private rented accommodation' (BEIS, 2021, p. 22), discussed in more detail below.

The final policy approach worth highlighting though not an energy demand policy *per se* is the implementation of smart meters. For the UK, this constitutes a key area of policy targeted at the domestic sector and includes claims and ambitions relating to energy demand reduction and fuel poverty. The Smart Metering Implementation Programme (SMIP) lays the legal foundation for energy companies to place a smart meter in every home and business by 2020. This flagship energy policy includes provision of smart meters with In-Home-Displays (IHDs) by energy companies to consenting householders (or business owners). It is intended to facilitate multiple other changes to the energy system (such as demand side management) but of interest for present purposes are the claims relating to domestic energy demand issues.

In this respect, it has been cast as a behavioural change programme that will deliver between 5 and 15% reductions in household energy consumption (Sovacool et al., 2017) and an opportunity to address issues of energy deprivation through easier switching, accurate billing, and more information to enable efficient usage, reducing bills and preventing debt accumulation (Gov.UK, 2018; Sovacool et al., 2017). The SMIP process and the claims about what it will deliver in terms of energy demand issues have, however, been subject to a series of controversies and problems in the process of roll-out. These have included socio-technical issues, such as installation failures, problems with the functioning of meters, and relatively low reported usage (60%) of the IHDs, which are central to claims regarding energy demand reduction and greater control for consumers (Sovacool et al., 2017).

Finally, I include within consideration of 'energy demand' policy, relevant UK transport policies. Though travel (or mobilities) is recognised within research as an equally important part of energy demand and an area of energy poverty, it is addressed almost entirely separately in UK policy through the Department for Transport (DfT). Arguably transport is an area of energy demand most in need of policy attention as there has been no net reduction in overall carbon emissions in this sector between 1990 and 2017 (Anable & Goodwin, 2019). Here, the current policy landscape focuses primarily on shifts to ultra-low and zero emission vehicles (e.g. electric cars) and efficiency gains as a way of reducing the carbon intensity of private vehicles, air travel, and haulage. There is very little in terms of policy that seeks to address current patterns in travel demand toward increasing growth, such as promoting alternative lower carbon modalities or more fundamentally seeking to alter the imperatives leading to increasing travel demands.

Anable and Goodwin (2019) mount a detailed critique of the assumptions underpinning current policy approaches to demand within the transport sector. They point to recent trends in car travel that have seen reductions in demand amongst age groups up to 59 and only a small increase in trips in the over 60s group. Using this evidence, they argue that the Department for Transport's current policy approach focusing on electrification should be replaced by policies that seek to lock-in, enhance, and further support these existing trends toward reduced car travel. They note that these trends have been shaped by policy in other areas aimed at reducing air pollution or enhancing health and quality of life, but that other policies can equally contribute to increasing demand (such as planning and development or centralisation of core services). They conclude with several recommendations for future policy, many of which include reference to the need to appraise the impact of non-transport policies on travel.

Overall, the changes in government policy since 2010 have seen policies aimed at energy demand reduction either cut or shifted in focus toward addressing fuel poverty. The emphasis continues to be on building efficiency as the primary issue of concern, with little space for more expansive understandings focused on questions about the creation of energy demand. At the same time, transport policy has failed to move beyond strategies that promote replacement of personal vehicles with low and ultra-low emission vehicles. Again, there has been little emphasis on reducing demand for car travel despite existing trends toward fewer trips (Anable & Goodwin, 2019).

This raises questions about the place for energy demand policy in the moves toward a net zero carbon target, which as noted above is an important, if not essential, area of policy for reaching those targets (Eyre & Killip, 2019). Such insights indicate a significant gap between the policy response and the findings currently dominating academic research, which suggest a need for an urgent and wide-ranging strategy for reducing energy demand across different areas and sectors. Though, within this, fuel poverty is being afforded more much needed policy attention, the strategies to address it are arguably constrained by narrow problem framings and understandings. The next section moves to consider wider fuel poverty policy and brings focus on the shifting definition of fuel poverty in the UK that underpins ECO. The following discussion thus brings to light questions about the framing of fuel poverty and the relevance of definitional struggles within this policy area.

Energy Poverty Policy in Focus

The shifts in energy demand policy that have seen the Energy Company Obligation (ECO) effectively become a fuel poverty scheme have been accompanied by ongoing debates about how to define, measure, and therefore identify the subjects to which the policy should be targeted. These definitional debates are important not least because they shape framings and understandings of the problem and consequently the nature of policy designed to address it. This section thus turns to address the way fuel poverty has come to be defined within policy highlighting academic critiques and exploring the importance of these debates to current strategies.

In the UK, fuel poverty came to be positioned as an urgent political problem owing, in part, to key statistical data through which the idea of 'excess winter deaths' was brought into being (Sovacool, 2015). The excess winter deaths figure denotes the number of additional deaths that occur during periods of cold weather and, in the UK, was pivotal in bringing fuel poverty to the attention of governing bodies. This brought with it a focus on older people and heat as defining features of both the problem and attempts to address it. Though fuel poverty had been defined and secured a place on the policy agenda from the 1980s onwards, it was Brenda Boardman's (1991) book that set out a specific definition, which has been utilised across the UK until recently.

Boardman (1991) defined fuel poverty as households whose fuel expenditure on all energy services exceeded 10% of their income. This was selected as a threshold because it represented twice the median expenditure on fuel and was the amount the poorest 30% of households were spending at that time (Moore, 2012). In terms of policy specifics, the expenditure threshold was modelled based on 'needed' energy consumption, rather than actual energy use. This was to avoid problems associated with genuinely fuel poor being missed due to underuse and wealthy but high energy users being included (Hills, 2012). 'Required' fuel costs were defined in terms of the fuel costs needed by a household to achieve a level of thermal comfort, adequate lighting, cooking, and typical appliance use to safeguard health (Moore, 2012). Under this definition, a core policy aim was established to *eradicate* fuel poverty by 2016.

The 10% definition was utilised to shape energy efficiency policies across the UK until England (though not the other UK nations[1]) adopted a new definition following a review (Hills, 2012). The new measure has been termed Low Income High Cost (LIHC) and defines fuel poverty as households that have a lower-than-average income and higher-than-average fuel costs. The calculation of higher-than-average fuel costs is based on modelled estimates of required (rather than actual) spending on domestic energy. The LIHC combines this estimate of fuel costs with income data that are used to identify households below a critical threshold income (i.e. less than 60% of the median income after housing and fuel costs). Any households that are below this level with higher-than-average fuel costs are defined as being in fuel poverty. The new measure brings an additional further emphasis on the *depth* of fuel poverty by reporting on the gap between the modelled required expenditure on fuel costs and actual expenditure—the so-called fuel poverty gap. This approach to measuring fuel poverty informs the delivery and assessment of policies like ECO and embeds particular ways of understanding the problem and its solutions.

The change in the definition has been critiqued with analysts such as Middlemiss (2016) highlighting some of its problematic effects. The first being that it further entrenches a distinction between fuel poverty and wider poverty and obfuscates links to wider structural causes, placing emphasis instead on energy efficiency as the primary policy response. The second problem, she argues, is that it concentrates attention on those who are deemed as being *in greatest need*, inferring a 'tacit acceptance' that some fuel poor households will be left outside of the remit. Finally, she points out that it decreases any recognition of the effects of changing prices in producing fuel poverty, further entrenching the idea that energy market reform is not one of the policy options relevant to energy poverty. This last point is perhaps particularly salient given high prices rises in recent years related, in part, to problems with the ways the energy market operates.

This change in definition has seen a shift in the policy problematisation of fuel poverty away 'from a condition that should and can be eradicated (as in the previous fuel poverty target), to a condition that can at best be alleviated' (Middlemiss, 2016, p. 2). The change in definition has had implications, then, for the high-level policy targets that have been set. Within the 2015 fuel poverty strategy (DECC/BEIS, 2015), the core target, underpinned by the LIHC measure, was to ensure that 'as many fuel poor homes as is reasonably practicable achieve a minimum energy efficiency rating of band C, by 2030 (DECC/BEIS, 2015, p. 20). This is a notably more tightly specified target than that of eradication and it has focused attention on a narrower set of issues related to the best way to achieve efficiency improvements.

For example, there are interim targets that include reaching a minimum rating of band E by 2020 and band D by 2025. This approach of setting interim targets has been critiqued as being unlikely to support meeting the 2030 target. Mainly because, in practice, this is likely to mean multiple stages of intervention and building work on homes to bring them up to the band C requirement over the period up to 2030, where a one-off whole home approach to energy efficiency is likely to be more effective in the long-term. This is an important point of critique, but it sits within a very narrow frame of what addressing fuel poverty means by focusing on the specifics of achieving energy efficiency, rather than giving room to the multiple ways in which energy vulnerabilities and precarity are constituted and capabilities undermined (see Bouzarovski & Petrova, 2015; Day et al., 2016; Middlemiss et al., 2019; Petrova, 2018).

Following consultation in 2019, BEIS published a new strategy for Sustainable Warmth addressing some of the critiques of the LIHC definition and associated approaches (BEIS, 2019, 2021). This new strategy includes an increased focus on a 'straight to C' approach where multiple measures for improving energy efficiency would be delivered to homes at one time, indicating a shift back to a whole house approach (BIES,

2021). It also includes changes to the measure of fuel poverty. This remains rooted in the principles of the LIHC measure but introduces a new absolute (as opposed to relative) high-cost threshold based on the energy efficiency rating of properties. Effectively anyone living in a home with an energy efficiency rating of D, E, F, or G would meet the high-cost criteria. The income threshold remains unchanged—any household with a residual income (after housing costs and energy needs) that would be below the poverty line are classed as meeting the income threshold. And a further significant change is that benefits targeted specifically at disability (Disability Living Allowance, Personal Independence Payment, and Attendance Allowance) are no longer counted as part of 'residual' or disposable income (BEIS, 2021, p. 11).

The new measure is called the Low-Income Low Energy Efficiency (LILEE) measure. As such, it is clear that the focus remains very much on housing efficiency. This is further emphasised in the ways that the fuel poverty gap will be assessed; i.e. by simulating the effect energy improvements would have on fuel costs and calculating the difference between a household's current costs and the simulated costs (BEIS, 2019). This is hoped to ensure stronger connection between the measurement of fuel poverty and its drivers, i.e. energy efficiency, fuel costs, and incomes. It is expected to account for; (1) improvements to energy efficiency taking people out of fuel poverty; (2) increases in fuel costs pushing people into fuel poverty; and (3) income inequality increasing the number of people in fuel poverty. Additionally, the changes to the measure are expected to make it less susceptible to 'churn'—i.e. people moving in and out of fuel poverty owing to the measure methodology, rather than any real change in their circumstance.

Beyond this, there are moves to bring greater emphasis on people deemed particularly vulnerable to cold (e.g. those with long-term severe physical and mental health problems or disabilities, very young children, and older people) signalling a continuing focus on energy for heating and particular groups that have historically been the target of interventions. In this, vulnerability is very much defined in terms of health and wellbeing, rather than wider issues of social participation, which are in themselves increasingly linked to health problems (e.g. Courtin & Knapp, 2017 on loneliness and health). The changes aim to ensure consideration of the impact of policies on the health and wellbeing of people on very low incomes even when they fall outside of the proposed fuel poverty metric (BEIS, 2019, 2021). This includes continuing possibilities for local government and other agencies to access policy measures and identify vulnerable households, without needing to adhere to the national measure.

All this gives an indication of the nature of the debate about fuel poverty in the UK policy context and the kinds of issues that are seen as within the scope of this policy area. Crucially, a focus on heat and energy efficiency remain central and shape a lot of the thinking and policy effort. However, even amidst this there is recognition of the limits of the measure and the need for local government and other agencies to be able to address fuel poverty in other forms, not only those defined nationally. Local and regional government bodies, charitable bodies, and non-governmental organisations both locally and nationally support people experiencing fuel poverty in a range of ways. These form part of the governance arrangements for energy poverty, while sitting outside of formal government.

Such activities include: (1) frontline referrals to the support and services available through national policy and wider central governance (such as ECO—meaning that it is not only the centralised application of the fuel poverty measure that is utilised to identify fuel poor); (2) use of local councils' crisis funds to support people experiencing issues related to fuel poverty; (3) local direct delivery of support for energy efficiency measures; (4) housing association support for home heating; (5) help with managing budgets and energy-related debt; and (6) legal advice (e.g. in the event of disconnection). It is at this scale that the lived experience of fuel poverty is best understood and where there is arguably most scope for application of wider understandings of fuel poverty as a multi-faceted issue that intersects with other policy areas.

Such possibilities for a wider understanding of the issues relevant to fuel poverty are also set in train by the introduction of a new sustainability principle within the Sustainable Warmth strategy (BEIS, 2021). This represents one of the more significant changes to the existing principles as it embeds an aim to take-into-account other government priorities in the delivery of policy to address fuel poverty. Though this is primarily focused on environmental policy goals, such as clean air and low carbon transitions, it does denote the likelihood of an end to existing fuel poverty policies that further entrench high carbon dependencies, such as extension of the gas grid. This signals a potentially important shift in the interconnections between fuel poverty policy and other energy demand policies. It also includes ambitions for connection to policy areas outside of energy, with specific focus on health policy—though this perhaps offers further openings for consideration of other policy areas too.

Despite these higher-level shifts in definitions, strategy, and the associated targets, the policies themselves have seen relatively little change in terms of focus and approach over several years. The energy efficiency policies, such as ECO (discussed above), have changed in their detail but have also remained consistent in terms of addressing the thermal and wider energy efficiency of homes with varying degrees of focus on low-income households and deprived areas (see Sovacool, 2015 for discussion of older policies not discussed here).

Beyond energy efficiency, there are a set of wider core fuel poverty policies in the UK, and these too have remained fairly constant over time. Among these are two key policies delivered as part of welfare benefits and funded through general taxation. These are the *Winter Fuel Payments*—a universal payment of £100–300 made to all households with a member over the age of 65—and the *Cold Weather Payments*—a payment of £25 per week made to households in receipt of some state benefits when the temperature drops below 0 degrees Celsius for more than seven consecutive days. A further notable policy is funded through the energy market by additions to energy bills. The Warm Home Discount Scheme—which is a discount for those on low incomes of £140 off household energy bills (given directly through the supplier) between September and March.

The introduction of minimum efficiency standards for the private rented sector also discussed above has been given greater prominence within fuel poverty policy with recognition of the high percentage (33.6%) of fuel poor households in the sector. While at present the requirement is only for landlords to bring houses up to an EPC (Energy Performance Certificate) rating of band E, ongoing consultation processes look set to strengthen these regulations, as well as improve practices for enforcement (BEIS, 2021). For example, renewed attention is being given to the Housing Health and Safety Rating System for its potential role in addressing fuel poverty. This mechanism is currently used by local councils to ensure that homes within the private rented sector are not hazardous (BEIS, 2021).

There have also been some wider policies that target pricing that although not always aimed directly at the fuel poor are in some way part of efforts to address energy poverty. Most recently, price caps have been introduced for people on default tariffs and those with prepayment meters to protect customers who tend not to (or cannot) switch suppliers (Ofgem, 2019). This approach has incurred criticism, however, amid wholesale gas price rises (see Bradshaw, 2021 for discussion) that have seen several smaller energy companies collapse, which they argue is partly because they lack capacity to pass rising costs on to consumers. The high gas prices in the UK have contributed to decisions to raise the energy price cap, with detrimental consequences for consumers. Previous policies in this vein tended to be more targeted and tackled pricing through mechanisms such as social tariffs, which offered discounted price plans for vulnerable customers and fuel poor households.

Smaller scale and trial policies have also been developed but ultimately not rolled out at scale. These include things like health services making referrals for energy efficiency measures and fuel poverty support, such as the 'Boilers on Prescription' project, which was trialed in Sunderland in the early 2010s. These schemes were evaluated and generally regarded as successful from their trials but not taken up as central government policy. There are, however, some ongoing attempts by wider agencies such as Citizens Advice (a charitable body that assists people with money, legal, consumer and wider problems) to support this kind of work, for example through the development of tool kits (e.g. see Citizens Advice, 2018).

The energy companies also have social responsibilities, in part, connected to the regulatory system, which to varying extents can be utilised to help fuel poor households and limit exposure to severe energy poverty for vulnerable groups. For example, energy companies cannot disconnect pensioners living alone or people living with children under 5 years of age between October and March (i.e. the winter months) regardless of payment arrears. The six largest suppliers (known in the UK as 'the big six') have also signed up to an agreement not to disconnect people at any time of year if they have a disability, long-term health problems, severe financial problems, or young children living at home. There is a further requirement on energy suppliers to offer debt payment plans for those in arrears. And voluntary social or corporate responsibility schemes sometimes offer financial support for customers, with such funds often being channelled into debt relief. Though this, in effect, amounts to energy companies using their social responsibility funds to pay themselves money owed.

Overall, the existing policy portfolio on fuel poverty has been critiqued for a continuing emphasis on older people, heat, and efficiency, with these concerns continuing to dominate policy mechanisms and older people receiving the majority of funding and protections (primarily owing to the

Winter Fuel Payment) (e.g. Middlemiss, 2016; Petrova, 2018; Simcock et al., 2016). Several studies have highlighted the energy precarity and vulnerabilities of other groups, such as young people, and the importance of a wider range of energy uses (beyond heat) to such groups (e.g. use of digital technology amongst young people) (see Middlemiss & Gillard, 2015; Petrova, 2018; Simcock et al., 2016). Analyses such as these and others (e.g. Bouzarovski & Petrova, 2015; Day et al., 2016; Middlemiss et al., 2019) seek to open-up narrow definitions and understandings of fuel poverty, bringing attention to a need for more multi-faceted innovative approaches that are grounded in lived experiences of fuel poverty and addressed at supporting the capabilities that energy services afford people.

Such analyses have potential to take thinking about energy poverty beyond domestic contexts and draw mobilities into the frame (see Chapter 2 for discussion). This brings the relevance of transport poverty policies into focus. Mattioli et al. (2017) identify a handful of policies that have implications for transport poverty even if not explicitly cast in these terms. Policies include: (1) the English National Concessionary Bus Travel Scheme—which offers free off-peak bus travel for English pensioners (i.e. the over 65s) regardless of income; (2) transport vouchers and other forms of financial support provided through the welfare state (often delivered at local levels, rather than national scale policy); and (3) housing densification policies that reduce car dependence and the need to travel long distances (even if this is not their aim). Despite this small range of policies, there is no explicit policy remit relating to transport poverty in the same way as that which exists for domestic fuel poverty. But this is an important area of critique given the clear role of access to transport in ensuring capabilities relevant to wellbeing and social participation, and the interconnections between fuel and transport poverty within lived experience. Though this is generally recognised within the energy poverty literature, little research has explored these linkages in depth.

The policy context for fuel poverty, then, remains one dominated by a focus on domestic heat and provision of efficiency measures with mobilities sitting outside of existing remits. Though domestic efficiency programmes undoubtedly have a role to play in addressing fuel poverty, research and theory are indicative of a far wider ranging set of concerns that shape energy vulnerabilities and precarity, including the move to encompass mobilities (Bouzarovski & Petrova, 2015; Day et al., 2016; Mattioli et al., 2017; Middlemiss & Gillard, 2015; Middlemiss et al., 2019; Petrova, 2018; Robinson & Mattioli, 2020; Simcock et al., 2016).

This literature highlights the importance of turning attention to the capabilities that energy services support and bringing into view the multiple policies, processes, and issues that shape experiences of energy poverty (Day et al., 2016; Middlemiss et al., 2019; Petrova, 2018). Such an approach to energy poverty that builds from lived experiences brings multiple different policy approaches (and areas) into view and provides footing for a wider dialogue about the issue and its solutions.

The linkages to areas outside of energy policy for understanding energy poverty challenges and issues have begun to be addressed in the academic literature with analysts such as Middlemiss (2016) and Snell et al. (2015) looking at welfare policy in particular. With the focus of the research in this book on welfare policy as an area of so-called invisible energy policy (i.e. non-energy policies that have effects for energy issues), it is to this area of policy that I now turn. In the remainder of this chapter, I focus on introducing the policy area of welfare and the remit and reforms that characterise the landscape in the contemporary UK context, before moving to discuss some existing intersections across welfare and energy policy.

Introducing Welfare as Invisible Energy Policy

Welfare and employment policy has a long history as a core part of governance arrangements in the UK. With its roots in the post-war reform periods of the late nineteenth and early twentieth centuries (Fraser, 1984), it is a policy area that has been important to the configuration of contemporary life. It is also a politically contentious policy area that has seen significant and often controversial changes to its remit and policies. In the UK, since 2010 there have been major reforms applied to the welfare system. These reforms have been enacted under the auspices of austerity (Pemberton et al., 2015) and entail both cuts to provision and changes that affect or restrict eligibility to receive benefits. This makes contemporary welfare policy a fast changing and politically contentious area of policy that provides scope for examining the impacts of policy change as they unfold and raises important questions about how such reforms may intersect with energy poverty.

In the UK, welfare and employment policy is delivered through the Department for Work and Pensions (DWP). The central thematic policy areas of this department are related to pensions and ageing, poverty and social justice, employment, and welfare policy and reform. Since the 2010 general election (which saw a Conservative led government take power from a previous Labour government), DWP has implemented rapid and controversial reforms principally focused on delivering cuts and changes to welfare provision for all *working-age* people in receipt of welfare assistance. The reforms have built on the existing approach to welfare implemented by the previous Labour government in focusing on 'getting people back into work' but have extended conditionality to all but the most severely disabled and have been far more punitive with the introduction of harsh sanctions (Wright, 2016).

These reforms encompass: the imposition of work coaches to advise working-age welfare recipients how to address perceived individual deficits (in, for example, their skills or work experience); new sanctions for those who fail to meet the requirements of their Claimant Commitment (e.g. missing an appointment with your work coach), including the partial and total withholding of their entitlements; new 'fit for work' assessments (Work Capability Assessments) for those in receipt of disability benefits (including those with physical disabilities, mental health and well-being problems, and those with chronic conditions); changes to disability benefits (e.g. changes incorporated in a shift from Disability Living Allowance—DLA—to Personal Independence Payments—PIP) that have effectively entailed cuts as well as redefinitions of eligibility; and changes to housing benefits including the under-occupancy charge (known as the 'bedroom tax' or 'spare room subsidy'), whereby if a home is considered too large for the occupant(s) then their social housing entitlement is reduced (Gov.UK, 2021; Roulstone, 2015).

All these changes have been applied in concert with a fundamental reform to the delivery of benefits—the shift to Universal Credit. Under the auspices of simplification, Universal credit involves the streamlining of multiple benefits into one single payment, encompassing and amalgamating Employment Support Allowance [ESA] for people with disabilities and ill health; Job Seekers Allowance [JSA]; income support; working tax credit; and housing benefit, amongst others. There have also been changes in the way the payment is received. Centrally, housing benefits now go directly to the claimant instead of being paid to the landlord, and payments are made monthly, as opposed to weekly or fortnightly.

These changes are meant to encourage better financial management and familiarity with payment patterns similar to many forms of employment. However, they have been applied in conjunction with a series of cuts to the amounts people receive placing increasing strain on people's ability to manage financially. The change in the regularity of payments has also had implications for people's ability to cope with very low incomes. Where a weekly payment might mean a person has limited funds (or none-at-all) for a few days, monthly payments can see people without money for a week or more. Moves to pay housing benefits directly to recipients, instead of landlords, at the same time as cuts have been applied has also seen some fall into arrears with rent, as they use the money to top up the deficits in new lower payments. The cost-cutting agenda has been further implemented through reductions in physical infrastructure, including closures of multiple job centres and shifts to move the benefits system online (Butler et al., 2018). It is important to note that through all these reforms older people aged over 65 have been protected from cuts, in terms of both state pensions and other benefits.

The wider political climate through this period has been a volatile one with multiple other political agendas intersecting with these changes. Not least severe and ongoing cuts to local government budgets and wider social care and health care services, such as the National Health Service (Duffy, 2013). There have also been changes in employment regulations that have seen increases in so-called zero hours contracts and other forms of insecure employment. This means that those no longer able to benefit from welfare systems can often move into poorly paid, insecure work. Though these policies are not the core focus for this book they remain relevant as wider context for the governance processes I address.

Contemporary welfare reform has been the focus of critique both within and beyond academia. One key area of critique has focused on the ways that such reforms 'have increasingly sought to reduce entitlements and intensify conditionality by individualising responsibility and mandating behaviour change' (Wright, 2016, p. 235). Wright (2016) highlights how these approaches are premised on 'a dominant model' of the welfare recipient as a 'unitary rational individual' that is personally responsible for their adverse circumstances and for taking action to resolve them (ibid.). Such models have been widely critiqued for failing to resonate with the lived experience of welfare recipients and removing policy focus from the wider structural issues that are constitutive of unemployment and poverty (e.g. Pemberton et al., 2015; Wiggan, 2012; Wright, 2016).

Other areas of critique relate to the treatment of particular groups under welfare reform, such as young people (under 25) and disabled people. For example, Roulstone (2015) points to a discursive shift in the construction of sick and disabled people toward dependency narratives and a corresponding retraction in related welfare spend. Austerity driven politics have seen the emergence of a discourse of 'focusing support on those with the greatest needs' (McVey, 2012—Parliamentary Under-Secretary for DWP 2012–2013—cited in Roulstone, 2015, p. 681). This has meant a significant reduction in welfare entitlements, with a 20% decrease in the number of recipients of Disability Living Allowance (now PIP) (Snell et al., 2015; see also Duffy, 2013).

In 2017, the UN Committee on the Rights of Persons with Disabilities published a report on the impact of UK Government policies on the rights of disabled people since 2010, finding that reforms had led to 'grave and systematic' violations of the rights of disabled people (Parliament UK, 2018). These findings have been the subject of vociferous dispute, with the UK government strongly disagreeing with the conclusions reached by others on the consequences of reform. Nevertheless, the reforms have been highly controversial and the subject of significant and sustained critical review.

These reforms to welfare occur within a context of wider ongoing programmes that extend across government but which also form a major part of changes to the ways that welfare systems operate. Not least of these are the moves to digitalisation that are being actively enacted and promoted by government through multiple policy departments. DWP Digital is an arm of the Department for Work and Pensions that operates to move the benefits system online embedding requirements for digital access within the welfare system (Butler et al., 2018). Though not a policy *per se*, this wider set of changes has major implications for accessing the benefits system, receiving payments, and meeting the new requirements for recipients (such as claimant commitments).

Beyond the specifics of welfare reform, welfare and employment policy has roles in other important areas, including, for example, housing policy. Welfare policy has had an important historic and ongoing role in the shifting nature of housing development. While there was a long-established tradition of government owned and rented housing, motivated in large part by concerns about population welfare, from the 1980s onwards these trends were dramatically changed to see levels of social and council housing reduce and private ownership sharply increase (see Butler et al., 2018). This was in no small part heralded by a welfare policy called 'The Right-to-Buy' wherein people were supported and encouraged to purchase their government-owned homes.

This policy continues today and is combined with other policies related to housing, such as housing benefit, which has been affected by both cuts, more stringent requirements (e.g. relating to the under-occupancy charge), and changes in the delivery of payments so that they no longer go directly to landlords. Added to this are online systems through which housing can be secured that rely on the individual to negotiate access themselves. Such changes combine to create a situation where welfare recipients are more likely to live in private rented accommodation in areas away from major service centres where cheaper housing is available. Housing policy changes, both historical and contemporary, have thus had major implications for the living situations of those in receipt of benefits.

The invisible energy policy literature has brought focus on how areas of policy, such as welfare, that seemingly have little connection to energy issues and no explicit remit in terms of energy can nonetheless have implications for energy demand concerns. Typically research in this space has focused on how non-energy policy areas affect sustainability issues related to energy demand (Gormally et al., 2019; Royston et al., 2018), but as I've asserted earlier in this book there is scope to direct analysis toward energy poverty and examine the ways that this issue is shaped by wider policy areas too (Cox et al., 2019). While no studies explicitly focusing on invisible energy policy have looked at energy poverty, there are a small number of articles that have focused on the intersections between welfare policy and fuel poverty. The discussion now turns to these studies along with an exploration of the ways that policies across welfare and fuel poverty currently intersect.

Interactions Across Energy and Welfare Policy

Clearly welfare policy has an important role in shaping experiences of poverty and consequently fuel poverty. In this respect, there are obvious connections across the two areas. This, however, is not reflected in how the different government departments or areas work together. Though there are some existing connections these are primarily related to technical aspects of policy delivery.

The Department for Work and Pensions (DWP) delivers the winter fuel and cold weather payments but within DWP they are viewed less as fuel poverty policy and more as welfare benefits. In addition to this, there is a connection in terms of data sharing and matching. For example, DWP provides income data that is used within the Low Income High Cost and

Low Income Low Energy Efficiency measures of fuel poverty. This data sharing role is, however, one that DWP undertakes for multiple government departments, as it holds the biggest repository of government social data. The connections, then, are limited and do not, at present, amount to collaboration on issues at the intersection of poverty and fuel poverty.

The limited nature of the connections is also reflected in the most recent strategy on fuel poverty policy in the UK (BEIS, 2019, 2021), wherein a new principle of ensuring compatibility between different policy aims considers low carbon policy and health but makes no mention of welfare policy. Partly, such lack of interconnection is undoubtedly related to the different cultures that shape the different departments. One—energy—is largely technical and dominated by market-based policies delivered through the energy companies and supported by levies on consumer bills. The other—welfare—is far more political and oversees policies that are funded through general taxation. Another related factor in this, though, is the long-standing battle to have fuel poverty recognised as something distinctive from wider poverty. And there are good reasons for wanting there to be a distinction.

In a key paper that looks across welfare and fuel poverty policy areas, Middlemiss (2016) has highlighted how the different political contexts within which these areas of policy are addressed often sees those in poverty demonised or positioned as undeserving of support, while those in fuel poverty do not face the same kind of vitriolic public discourse. For many, distinguishing fuel poverty as a separate issue has been achieved through a long fought battle and the benefits of it being treated as separate far outweigh the problems. Some, however, have pointed to issues with this distinction arguing that it obscures structural causes of fuel poverty and entrenches a focus on energy efficiency over and above measures that address income inequality, energy market reform, or the cost of living and energy prices (Middlemiss, 2016; Petrova, 2018).

Middlemiss (2016) argues that the focus on energy efficiency obscures the lived experience of energy poverty and the complex interrelated 'assemblages' shaping such experiences, including poor health, major life events such as bereavement, employment, housing situation, and so on (also see Harrison & Popke, 2011; Middlemiss & Gillard, 2015). She highlights how channelling policy attention in this way means that those on low incomes but living in energy-efficient homes are overlooked even if they still cannot afford sufficient energy services necessary to support health, wellbeing, and social participation. She links this to differences in the characterisation of the subjects of these different areas of policy, noted above, and argues that a considerable distance has been constructed between the two policy areas despite addressing a very similar population. Where in the case of fuel poverty the focus is on technical solutions and infrastructural efficiencies, in welfare the emphasis is placed on individual behavioural problems and deficits e.g. in skills and/or motivation (Wright, 2016).

These differences in political positioning of the subjects of welfare and energy poverty policy make it more difficult to imagine collaboration between the different departments. The possibilities for collaboration and forms of cross departmental governance in this case, thus, appear constrained. But there are good reasons for thinking about the different policy areas in combination and understanding the implications of welfare policy for energy poverty. For example, in another notable paper on this topic, Snell et al. (2015) show how disabled people have simultaneously been affected by reduced incomes owing to welfare reform, *and* changes in their eligibility for fuel poverty support (i.e. under the LIHC/LILEE measures), despite evidence showing the increased vulnerability of disabled people to fuel poverty.

In sum, there are clear connections between the issues of fuel poverty and wider poverty but there are also stark divisions in how these issues are understood and addressed through policy. There are also other welfare and employment policies and processes that can be thought about in terms of their implications for energy poverty. These include housing policies and benefits, for example, that have seen changes affecting the nature of property ownership, the dynamics of where people live, and the types of housing in which they are likely to live. Or the moves to digitalisation of the benefits system, which further intertwines with other cuts and changes to have implications for issues of energy poverty. While welfare reforms have been given some attention in terms of energy poverty within the notable contributions discussed here, the wider policy area of welfare taking in multiple reforms and policy shifts along with their interconnections represents an important area for further analysis.

Conclusion

This chapter has examined the contemporary UK policy context for energy poverty and introduced the 'invisible-energy policy' case on which the empirical analysis within the remaining chapters is focused—namely welfare policy. I highlight how the policy understanding of fuel poverty is still very much (and increasingly) focused on heat and building efficiency, but I argue that there are some policies and approaches that create room for more complex understandings of energy poverty. These include recognition of the important role of local governing bodies and non-governmental organisations in identifying and supporting those experiencing energy poverty. Such modes of delivery for support can operate outside of the narrower framings embedded in the national fuel poverty measures and are therefore less restricted in both who they support and the approaches they adopt. This is a space in which more complex understandings of the dynamics constituting energy poverty are likely to have greater purchase.

The possibilities for and limits to synergies across policy areas have also been discussed. In this respect, I suggest that although there are severe constraints on policy remits and cultures, there are also some openings for connections to be made. The introduction of a sustainability principle to fuel poverty policy in the UK is interesting in this respect (BEIS, 2021). This has brought focus on connections across policy issues, such as air pollution and carbon emissions, but has also opened-up to other areas like health signalling further possibilities. The inclusion of a different approach to income in the new measures of fuel poverty (Low Income High Cost/Low Income Low Energy Efficiency) has also seen connections being forged between the departments responsible for welfare policy and energy poverty. Though, as noted, these connections are quite limited, they could offer avenues for new forms of dialogue to emerge.

Ultimately, I argue that looking outside of energy policy and exploring intersections across policy areas and issues makes possible a fuller engagement with alternative ways of conceptualising energy poverty. In this book, I work to develop the capabilities-based energy poverty frameworks (discussed in Chapter 2) in combination with ideas and insights founded in practice-based energy research (see Chapter 3) to build distinctive insights relevant to understanding the dynamics of energy poverty. The remaining chapters present an analysis of interviews undertaken with those implicated in welfare policy (see the introductory chapter for description of the methods and wider project). These chapters build insights relevant both to a deeper understanding of energy poverty and to the value of bringing these different conceptual traditions together. This entails advancing the flexibility inherent in capabilities and energy services-based perspectives by incorporating discussion of multiple energy services, across electrical technologies and heat, and through to mobilities.

While as discussed here there have been some analyses of the implications of welfare policy for energy poverty (e.g. Middlemiss, 2016; Snell et al., 2015), there are no studies that address the idea of invisible energy policy explicitly within this space. Nor are there analyses that draw together the wider conceptual insights developed by practice theory-inspired energy scholars with concepts for thinking about energy poverty. The chapters that follow offer a bounded analysis of data from a wider project focused on welfare policy as an area of invisible energy policy. The emphasis is on developing and opening-up analytic lines that work across the intersections of practice theory-based research (including invisible energy policy) and energy poverty concepts or insights (such as those relating to capabilities and precarity).

First, Chapter 5 is directed at examining the ways that invisible energy policy arising from welfare shapes the emergence of energy poverty with deep implications for capabilities. In this, the chapter explores the value of the invisible energy policy agenda for better understanding the emergence and experience of energy poverty. I argue that it can be utilised to extend analysis of policy areas and the wider discourses and narratives that define them (cf. Middlemiss, 2016), as well as facilitating an openness that can allow insights relevant to developing capabilities-based understandings of energy poverty. This chapter thus draws together ideas from across the practice theory-inspired invisible energy policy agenda (Cox et al., 2019; Royston et al., 2018), with those from energy poverty research including work on political problematisations (Middlemiss, 2016), precarity (Petrova, 2018), and energy capabilities (Day et al., 2016; Middlemiss et al., 2019). It sheds light both on the dynamics of energy poverty and on the ways that the invisible energy policy agenda can have value for examining issues of energy deprivation.

Following this, Chapter 6 gives attention to how ideas concerning the *constitution of need* from practice-based energy research (e.g. Shove et al., 2012) can be important for understanding energy poverty. Again, drawing on a capabilities-based understanding of energy poverty, this chapter develops insights into the ways that non-energy policy affects not only people's abilities to meet energy service needs, but their possibilities for shaping and resisting those needs too. This takes a cue from Walker's (2013) analysis where concepts of recruitment and defection as processes that shape the emergence and decline of different practices are explored for their relevance in thinking about inequality. But the analysis goes further by drawing in the invisible energy policy agenda to explore the ways that different areas of governance exert power. I argue that marginalisation, evident for those subject to some forms of welfare policy, constrains possibilities to resist or defect from practices, and that processes of constitution can be far more oppressive that is suggested by existing work exploring these ideas. This brings focus to how understandings of power and inequality can be brought more firmly into practice theory-based analysis and research on energy.

Note

1. The empirical research was undertaken in England. As such, I do not discuss the other nations within the UK where there are some marked differences in approach including retention of the 10% fuel poverty definition and eradication targets.

References

Anable, J., & Goodwin, P. (2019) Transport and mobility. In N. Eyre & G. Killip (Eds.), *Shifting the focus: Energy demand in a net-zero carbon UK*. Centre for Research into Energy Demand Solutions. ISBN: 978-1-913299-00-2

BEIS (Department for Business, Energy and Industrial Strategy). (2019). *Consultation on the fuel poverty strategy for England*. BEIS.

BEIS (Department for Business, Energy and Industrial Strategy). (2021). *Sustainable warmth: Protecting vulnerable households in England*. BEIS [online] Available at: https://www.gov.uk/government/publications/sustainable-warmth-protecting-vulnerable-households-in-england. Accessed 10 September 2021.

Boardman, B. (1991). *Fuel poverty: From cold homes to affordable warmth*. Belhaven Press.

Bouzarovski, S. (2018). *Energy poverty: (Dis)assembling Europe's infrastructural divide*. Palgrave.

Bouzarovski, S., & Petrova, S. (2015). A global perspective on domestic energy deprivation: Overcoming the energy poverty–fuel poverty binary. *Energy Research & Social Science, 10*, 31–40.

Bradshaw, M. (2021). *Gas price spike: How UK government failures made a global crisis worse*. [online] Geography Directions. Available at: https://

blog.geographydirections.com/2021/09/23/gas-price-spike-how-uk-government-failures-made-a-global-crisis-worse/. Accessed 6 January 2022.

Butler, C., Parkhill, K. A., & Luzecka, P. (2018). Rethinking energy demand governance: Exploring impact beyond "energy" policy. *Energy Research & Social Science, 36*, 70–78.

Citizen's Advice. (2018). *Building cold home referrals with the health sector, toolkit* [online]. Available at: chrome-extension: https://efaidnbmnnnibpcajpcglclefindmkaj/viewer.html?pdfurl=https%3A%2F%2Fwww.citizensadvice.org.uk%2FGlobal%2FCitizensAdvice%2FHealth%2520professionals%2520cold%2520homes%2520toolkit.pdf&clen=5936632&chunk=true. Accessed 10 September 2021.

Committee on Climate Change. (2019). *Reducing UK emissions—2019 progress report to parliament—committee on climate change* [online]. Available at: https://www.theccc.org.uk/publication/reducing-uk-emissions-2019-progress-report-to-parliament/. Accessed 10 September 2021.

Courtin, E., & Knapp, M. (2017). Social isolation, loneliness and health in old age: A scoping review. *Health and Social Care in the Community, 25*(3), 799–812.

Cox, E., Royston, S., & Selby, J. (2019). From exports to exercise: How non-energy policies affect energy systems. *Energy Research & Social Science, 55*, 179–188.

CSE (Centre for Sustainable Energy). (2014). *Evaluation of the ECO and green deal: Executive summary, centre for sustainable energy, Bristol* [online]. Available at: https://www.cse.org.uk/projects/view/1225. Accessed 10 September 2021.

Day, R., Walker, G., & Simcock, N. (2016). Conceptualising energy use and energy poverty using a capabilities framework. *Energy Policy, 93*, 255–264.

DECC/BEIS (Department of Energy and Climate Change/Department for Business, Energy and Industrial Strategy). (2015). *Fuel poverty strategy for England*. BEIS.

Duffy, S. (2013). *Centre for welfare reform: A fair society?* [online]. Available at: https://www.centreforwelfarereform.org/library/a-fair-society1.html. Accessed 10 September 2021.

Eyre, N., & Killip, G. (Eds.). (2019). *Shifting the focus: Energy demand in a net-zero carbon UK*. Centre for Research into Energy Demand Solutions. ISBN: 978-1-913299-00-2

Fraser, D. (1984). *The evolution of the British welfare state* (2nd ed.). Macmillan Press.

Gormally, A. M., O'Neill, K., Hazas, M. D., Bates, O. E. G., & Friday, A. J. (2019). "Doing good science": The impact of invisible energy policies on laboratory energy demand in higher education. *Energy Research & Social Science, 52*, 123–131.

GOV.UK. (2018). *Smart meters: A guide* [online]. Available at: https://www.gov.uk/guidance/smart-meters-how-they-work#benefits-of-smart-meters. Accessed 10 September 2021.

GOV.UK. (2021). *Universal credit* [online]. Available at: https://www.gov.uk/universal-credit/your-responsibilities.

Greene, M., & Fahy, F. (2020). Steering demand? Exploring the intersection of policy, practice and lives in energy systems change in Ireland. *Energy Research & Social Science, 61*, 101331.

Harrison, C., & Popke, J. (2011). "Because you got to have heat": The networked assemblage of energy poverty in eastern North Carolina. *Annals of the Association of American Geographers, 101*(4), 949–961.

Hills, J. (2012). *Getting the measure of fuel poverty. Final report of the fuel poverty review.* London: CASE report 72. [online] GOV.UK. Available at: https://www.gov.uk/government/publications/final-report-of-the-fuel-poverty-review. Accessed 5 January 2020.

Mattioli, G., Lucas, K., & Marsden, G. (2017). Transport poverty and fuel poverty in the UK: From analogy to comparison. *Transport Policy, 59*, 93–105.

Middlemiss, L. (2016). A critical analysis of the new politics of fuel poverty in England. *Critical Social Policy, 37*(3), 425–443.

Middlemiss, L., & Gillard, R. (2015). Fuel poverty from the bottom-up: Characterising household energy vulnerability through the lived experience of the fuel poor. *Energy Research & Social Science, 6*, 146–154.

Middlemiss, L., Ambrosio-Albalá, P., Emmel, N., Gillard, R., Gilbertson, J., Hargreaves, T., Mullen, C., Ryan, T., Snell, C., & Tod, A. (2019). Energy poverty and social relations: A capabilities approach. *Energy Research & Social Science, 55*, 227–235.

Moore, R. (2012). Definitions of fuel poverty: Implications for policy. *Energy Policy, 49*, 19–26.

Ofgem. (2019). *About energy price caps*. Available at: https://www.ofgem.gov.uk/energy-price-caps/about-energy-price-caps

Parliament UK. (2018). *UN committee report on the implementation of the convention on the rights of persons with disabilities: Debate on the government's response* [online]. Available at: https://researchbriefings.parliament.uk/ResearchBriefing/Summary/CDP-2018-0148. Accessed 10 September 2021.

Parliament UK. (2019). *Government gives details on setting a UK net zero emissions target*. Available at: https://www.parliament.uk/business/news/2019/june/government-gives-details-on-setting-a-uk-net-zero-emissions-target/. Accessed 10 September 2021.

Pemberton, S., Fahmy, E., Sutton, E., & Bell, K. (2015). Navigating the stigmatised identities of poverty in austere times: Resisting and responding to narratives of personal failure. *Critical Social Policy, 36*(1), 21–37.

Petrova, S. (2018). Encountering energy precarity: Geographies of fuel poverty among young adults in the UK. *Transactions of the Institute of British Geographers, 43*(1), 17–30.
Robinson, C., & Mattioli, G. (2020). Double energy vulnerability: Spatial intersections of domestic and transport energy poverty in England. *Energy Research & Social Science, 70*, 101699.
Roulstone, A. (2015). Personal Independence Payments, welfare reform and the shrinking disability category. *Disability & Society, 30*(5), 673–688.
Royston, S., Selby, J., & Shove, E. (2018). Invisible energy policies: A new agenda for energy demand reduction. *Energy Policy, 123*, 127–135.
Shove, E., Pantzar, M., & Watson, M. (2012). *The Dynamics of social practice: Everyday life and how it changes*. Sage.
Simcock, N., Walker, G., & Day, R. (2016). Fuel poverty in the UK: Beyond heating? *People Place and Policy Online, 10*(1), 25–41.
Snell, C., Bevan, M., & Thomson, H. (2015). Welfare reform, disabled people and fuel poverty. *Journal of Poverty and Social Justice, 23*(3), 229–244.
Sovacool, B. K. (2015). Fuel poverty, affordability, and energy justice in England: Policy insights from the Warm Front Program. *Energy, 93*, 361–371.
Sovacool, B. K., Kivimaa, P., Hielscher, S., & Jenkins, K. (2017). Vulnerability and resistance in the United Kingdom's smart meter transition. *Energy Policy, 109*, 767–781.
Walker, G. (2013). Inequality, sustainability, and capability: Locating justice in social practice. In E. Shove & N. Spurling (Eds.), *Sustainable practices: Social theory and climate change* (pp. 181–196). Routledge.
Wiggan, J. (2012). Telling stories of 21st century welfare: The UK Coalition government and the neo-liberal discourse of worklessness and dependency. *Critical Social Policy, 32*(3), 383–405.
Wright, S. (2016). Conceptualising the active welfare subject: Welfare reform in discourse, policy and lived experience. *Policy & Politics, 44*(2), 235–252.

CHAPTER 5

Invisible Energy Policy and Energy Capabilities

Abstract This chapter explores the value of bringing thought about invisible energy policy together with key analytic endeavours in the field of energy poverty. It uses empirical material to develop understanding of how capabilities that are linked to experiences of energy deprivation are shaped by (non-energy) policy. Within this, the chapter explores the potential for the invisible energy policy orientation to advance existing work related to the ways that wider discourses and framings shape experiences of energy poverty issues. The chapter gives particular focus to the implications of relations between discourses of fuel poverty and those of broader poverty, arising from energy and welfare policy, respectively, extending analysis by exploring how such discourses act upon subjects in ways that affect possibilities for challenging conditions of energy poverty.

Keywords Invisible energy policy · Energy poverty · Welfare policy · Energy precarity · Capabilities

Introduction

This chapter uses in-depth empirical data derived from interviews with people implicated in welfare policy (see Chapter 1 for details of the research) to explore and draw together different conceptual ideas from across energy demand research (see Chapters 2 and 3 for a discussion

C. Butler, *Energy Poverty, Practice, and Policy*, Progressive Energy Policy,
https://doi.org/10.1007/978-3-030-99432-7_5

of key concepts). Centrally, it examines how the practice theory-inspired invisible energy policy agenda (Cox et al., 2019; Royston et al., 2018) can be combined with key ideas from energy poverty research (Day et al., 2016; Middlemiss, 2016; Middlemiss et al., 2019; Petrova, 2018) to create distinctive insights into energy deprivation.

The chapter looks at how policies in the non-energy area of welfare directly shape energy deprivation by affecting access to energy services and related capabilities. This aligns with previous invisible energy policy research (see Cox et al., 2019) but brings greater focus on the value of this research agenda for energy poverty. However, the analysis also goes beyond this to suggest the reorientation of perspective offered by looking outside of energy policy brings insights important for understanding the dynamics of energy poverty amid wider forms governance. I build from the assertion that using analysis of welfare policy as a starting point takes one outside of that which would conventionally form the focus for looking at issues of energy deprivation. It gives a view of governance processes that orients analysis beyond the existing categories and structures of government and policy at the outset (see also Butler et al., 2018). Specifically, the analysis reflects on the ongoing concern with the links and disconnections between energy and poverty, or across fuel poverty and wider poverty, and the wider structural conditions that are implicated in energy deprivation (e.g. Bouzarovski, 2018; Middlemiss, 2016; Petrova, 2018).

The chapter thus builds to develop the invisible energy policy agenda by going beyond analysis of more direct forms of policy influence on energy poverty to examine *policy discourses* across energy and welfare policy (cf. Middlemiss, 2016) and, crucially, develops this by examining the ways they act upon people and shape practice. This analytic endeavour works to show how policy and political discourses shape experiences of energy poverty in fundamental ways, and it reveals what they obscure in terms of understanding and addressing energy poverty. Centrally, it demonstrates how welfare policy contributes to marginalisation in ways that are inextricably connected to, and foundational for, experiences of energy poverty and its normalisation as part of everyday practice.

The analysis adopts an approach to energy poverty that builds from and advances the capability-based frameworks discussed in Chapter 2, drawing in practice theory ideas relating to invisible energy policy (discussed in Chapter 3). This conception is put to work in ways that afford a flexible approach to energy poverty looking across multiple services and elements of practice. This includes those energy services conventionally addressed within policy (e.g. heat) but also those outside of current policy remits, such as mobilities and travel, or considered as only of minimal importance because of relatively low levels of direct energy requirement, such as information and communication technologies (ICTs). The chapter offers insight into the ways that the invisible energy policy agenda can have value for examining issues of energy deprivation and advances key lines of enquiry for understanding the dynamics of energy poverty.

Living with Energy Poverty

The analysis in this first section highlights implications of major welfare reforms for capabilities related to energy poverty and unpicks policy distinctions between fuel poverty and wider poverty (cf. Middlemiss, 2016) building insight into broader social and political processes that shape the issues. In the following discussion, all extracts are labelled to distinguish between the type of interviewee (i.e. *biographical* for those affected by the welfare system, *stakeholder* for those with professional roles in this area) and the location of interview (i.e. *Bristol* or *York* as the case study areas or *national* for those working at this scale). They are also numbered to allow for different interviewees to be identifiable and where quotes are from the workshops this detail is added to the descriptor. For more detail on the methods see Chapter 1.

As discussed in Chapter 4, the period since 2010 in the UK has seen major reforms related to welfare provision for working-age people. Key changes include: (1) the introduction of new conditions for receipt of benefits (e.g. claimant commitments) and associated sanctions if these are not met; (2) work capability assessments for those currently in receipt of disability benefits and related cuts; (3) the introduction of a new system for delivery of benefits—namely Universal Credit—again with associated cuts to benefits in real terms and new monthly payments replacing weekly or bi-weekly ones; and (4) changes to housing benefit involving the introduction of the under-occupancy charge (or the bedroom tax) being applied to people living in properties deemed as having more bedrooms than necessary. The research here highlights how these reforms have severely affected multiple aspects of life that have key implications for energy poverty and related capabilities. The analysis explores this with focus on revealing how current policy definitions restrict recognition of the ways that experiences of energy poverty are intimately bound up with those of wider poverty. The first quote below highlights severe domestic energy deprivation because of a person's benefits being stopped following a work capability assessment (WCA), in which the participant was deemed fit for work. The benefits were later reinstated after a lengthy appeal process through which the original WCA decision was overturned.

> *Interviewer*: May I ask how you coped over that year when they stopped your payments?
>
> *Interviewee*: With huge difficulty. Huge, huge difficulty. It was not an easy year at all. Definitely one of the hardest I've experienced. Yeah, very dark… it was a really bad situation…. In a lot of debt with bills and stuff… The power for the flat would turn off, sort of thing… I never had the heating on. Never, ever had the heating on. I only had it for hot water for showers…The flat was just horrendously damp… You know, blankets, all the rest of it. Just shiver. Yeah, it could get very cold in that flat…" (Biographical Interviewee 4, Bristol)

The prevalence of problems associated with incorrect assessments in welfare reform has been revealed by analysis of cases elsewhere across both academic research and news media (e.g. Duffy, 2014; Morris, 2013; Roulstone, 2015). The example here is to highlight the implications of such experiences for energy deprivation. The emphasis is on what one might think of as the conventional focus of fuel poverty policy—namely heat and electricity in the home—with these services either severely limited or lost entirely. This case brings to light the obvious and clear connections between policies that relate to poverty more widely and energy poverty. However, this clear overlap is more difficult to see or perhaps engage with from the technologically and efficiency-oriented perspectives that characterise much fuel poverty policy.

As discussed in detail in Chapter 4, though such policies utilise assessments of income they tend to emphasise building efficiency as the primary focus for resolution (also see Middlemiss, 2016). In this case, the house in which the participant lives had identifiable issues with building efficiency and quality (e.g. damp), but tackling this alone would not have addressed the problems the person faced in terms of energy deprivation and having basic capabilities, primarily because their income was at such a low level that building efficiency measures would still not have afforded them the capability to access adequate lighting and heat. Given this, the quote brings into focus an initial set of questions about the distinction between fuel poverty and wider poverty.

Centrally, in examining experiences like the one above some of the inherent problems with such distinctions become apparent. Clearly, the issue here was not one solely of energy efficiency or being able to afford heat but was bound up with their inability to pay for any basic daily needs for an extended period. This, in turn, resulted in mounting debt that left a legacy even when the person's benefits were eventually reinstated. In this example, then, the role of welfare policy in shaping energy deprivation is direct and clear but by virtue of a policy distinction, would not be considered as an example of fuel poverty; rather, this would simply be poverty.

Across the research, the inextricable links to wider poverty and the production of energy deprivation were clear (see also Mould & Baker, 2017). The next example concerns a participant who experienced six months of sanctions for failing to attend an appointment. She also was struggling to repay emergency loans provided through the welfare system and had become subject to the so-called bedroom tax following her partner's death (which resulted in her two-bedroom flat being reclassified as under-occupied).

> I can't afford to heat that [home] at all, no. I don't. Just put a quilt round me, dog's got a little blanket, can't afford it. The night storage heaters I can't afford to use…The cooker don't work so I've been without a cooker for the past three years so I've only got a toaster and a kettle, basic…. (Biographical Interviewee 8, Bristol)

This example goes beyond issues of energy affordability to highlight limited access to energy services related to materials and infrastructure but not confined to efficiency. The participant talks about their lack of basic appliances for cooking, as well as the specific form of heating system they have (electric storage heaters), which tends to be more expensive. A capability-based approach to energy poverty brings focus on how these conditions shape possibilities for access to food and mental health as well as the more often acknowledged implications of cold for physical health. The orientation of the research toward invisible energy policy brings focus onto the wider systemic processes and policies shaping these conditions. Noticing the impacts of welfare policy on energy poverty highlights challenges associated with both narrow definitions of the problem and policy solutions confined to energy efficiency.

This orientation also directs attention to more specific forms of cross over between policy areas that are otherwise not in view. For instance, existing fuel poverty policy only addresses problems with income through the winter fuel payments (for those in receipt of state pensions only) and the cold weather payments—the latter of which is allocated based on receipt of existing benefits (and other vulnerability characteristics). The advent of processes that stop people receiving benefits (such as capability assessments) has implications, then, not only for income more generally but also for access to the limited fuel poverty policies targeted at income issues. Such changes within welfare policy thus both compound issues of energy deprivation and affect access to forms of fuel poverty support.

The next example addresses wider dimensions of welfare policy—beyond eligibility assessments, sanctions, and cuts—that shape energy deprivation, focusing on policies that result in people being more likely to live in poor housing. Key dimensions of welfare policy related to housing, both historically and in the context of contemporary reforms, have had important implications for energy deprivation. Historically, major reforms to welfare policy have seen housing move from a position where it was predominantly built and owned by the state to one of private construction and ownership (see Butler et al., 2018), such that 'proper' housing consumption is now synonymous with home ownership (Petrova, 2018). These processes along with recent reforms related to the level of benefits for housing and the removal of payments direct to landlords (see Chapter 4 for discussion) have seen increasing concentrations of people on welfare in private rented sector accommodation (as opposed to council or social housing).

> ...people are being pushed more and more into bad landlords as they can't afford to live anywhere else.... (Stakeholder Interviewee 2, National Agency)

As highlighted in Chapter 4, private rented sector accommodation has often been neglected within fuel poverty and wider policy, with limited steps taken to regulate the sector and ensure housing meets efficiency standards. Even with contemporary regulations, significant problems remain in enforcement, with local councils being largely responsible but lacking in funding to support major programmes of action (BEIS, 2021). The research data were revealing not only in terms of the poor quality of housing experienced by the participants, but in terms of other dimensions of policy that restricted their ability to select appropriate or better-quality housing. The next extract speaks to how a combination of reforms and processes for accessing housing within welfare provisioning shapes not only experiences of domestic energy poverty (related to the quality of housing) but also forms of deprivation related to transport and mobilities (arising from the location of housing). In line with Middlemiss et al. (2019), the relations between capabilities and energy poverty are cast as complex with social relations (a basic capability) ultimately shaping the ability to negotiate better housing conditions within systems that tend not to support vulnerable people.

> Basically [Name], my housing [provider], they told me that this was my only option... When I look back on it now and the position I'm in, they weren't actually allowed to do that. I could have refused it. They would have had to offer me somewhere else. But at the time I was so vulnerable emotionally because I just wanted to get away from where I was and they just offered me this...I constantly have problems with them now, with my housing. (Biographical Interviewee 6, Bristol)

This participant experienced problems with their quality of housing, having issues with mould and damp, and their location, being distant from family, friends, their work and services. This underpinned a lack of adequate warmth and created requirements for travel but without the means or access to fulfil them. Bound up within this participant's narrative were multiple ways that these issues, related to energy services, shaped their capabilities, such as their ability to maintain social relations and to have mental and physical health. But as highlighted above their capabilities were also important in the processes through which these forms of energy deprivation were initially constituted (cf. Middlemiss et al. 2019). The extract above exposes two important ways that welfare policy shaped these experiences: first, it points to the relevance of constraints on people who are subject to welfare policy in terms of choice in housing; second, it highlights the ways that forms of capability are shaped by wider policies and structural processes to result in further marginalisation. In particular, this participant discusses their social relations and vulnerability at the time of negotiating new housing as affecting their allocation. Underlying all of this, of course, is the prevalence of poor and inadequate housing but I argue that the processes through which people in contexts of low income are pushed into inadequate housing are equally as important to examine. This resonates with Großmann et al.'s (2014) findings in the German context, where mechanisms of housing market discrimination and subsequent residential segregation were shown to have caused low-income households to live in low-quality housing. In later work, Großmann and Kahlheber (2017) argue that such processes remain largely invisible in the context of energy poverty research and policy, occluding the recognition of wider systemic processes in the constitution of energy poverty. By looking outside of energy policy, in this case to experiences of welfare policy, analysis is afforded greater possibilities for understanding the wider processes that underpin the creation of energy poverty.

All this undermines the idea that there is a wholly positive outcome of upholding a firm and clear distinction between energy poverty and wider poverty. However, it is not to say there is nothing distinctive about energy poverty—quite the contrary. This type of evidence also demonstrates the *essential* nature of energy in being able to enact practices and fulfil multiple basic capabilities, such as those related to living a healthy life, and if anything reinforces the importance of addressing it with targeted policies. However, it does also bring into sharp focus the insight that the causes of energy poverty could be better addressed both within energy policy and in wider policy through greater attentiveness to the interlinkages between issues.

This argument about interconnection can be taken further by looking at the ways that energy poverty further entrenches poverty more generally. In the quote below, the participant discusses challenges related to their low income that was destabilised as they transitioned to Universal Credit. The focus in this part of the narrative is on access to transport as an important energy service that supports multiple capabilities, in this case relating to the ability to secure income.

> I don't go out because I can't really go out... Say if you have meetings or appointments or like a job interview or whatever, it's the most embarrassing thing asking where it is or whatever and then having to walk all that way. (Biographical Interviewee 10, Bristol)

Here, the participant highlights the implications of energy deprivation for capabilities relating to social respect and access to work or income. This foregrounds the interrelations between energy poverty—in this instance related to mobilities—and wider poverty as access to work and income is restricted. Their ability to afford and use transport options was, in turn, shaped by changes within welfare policy that saw cuts to their income as well as changes to the timing of payments. In other work, Mattioli et al. (2017) have noted the recursive link between transport poverty and economic stress, highlighting issues such as car dependency required for access to work. The above quote foregrounds the role of transport poverty in affecting possibilities for accessing work at all through constraining abilities to participate in interviews and meet appointments. This signals a cyclical relationship between energy poverty and wider poverty highlighting again the importance of engaging with the relations between these issues. It is possible to see how energy poverty when understood in terms of capabilities—as opposed to building efficiency and income—can underpin and shape experiences of wider poverty in fundamental ways. As much as being a subject of welfare policy can have serious implications for energy deprivation, then, so can lack of access to energy services shape the reproduction of poverty.

In analysing the interlinkages between energy poverty and wider poverty, an important point has been articulated about the very different politics of fuel poverty policy as compared to welfare policy (which is meant to alleviate poverty) or indeed wider political discourse about poverty (Bouzarovski, 2018; Middlemiss, 2016). Middlemiss (2016) highlights how while welfare policy has long been entrenched in notions of deserving and undeserving subjects, of individualised causes characterised in terms of personal deficits (e.g. in willingness to work or skills) (e.g. see Butler et al., 2018; Pemberton et al., 2015), the subjects of fuel poverty have been cast in a very different blameless light where recipients are positioned as worthy of support and help.

This foregrounds a central challenge relating to the conclusions one might reach from the analysis here. While I have problematised the boundary between fuel poverty and wider poverty, the very distinctiveness which obscures the links between policy areas also creates room for both a governmental budget that does support those living in (fuel) poverty and a far less punitive discourse around the issues. However, the next section addresses questions about the differing discursive repertoires of those subject to both welfare policy and energy poverty and opens up the analysis of the fuel poverty/poverty distinction further. The research data suggest that such differences in the framing of fuel poverty and wider poverty do not necessarily translate into more positive experiences for those people that are subject to both narratives. The analysis highlights the institutionalisation and normalisation of practices that evolve from living without energy as an important way that political discourses and policies—beyond energy policy—shape experiences of fuel poverty.

Living Without Energy

There has been much discussion within policy and analysis of fuel poverty about issues of *self-disconnection* and the closely associated idea of *self-rationing* (e.g. Hargreaves & Longhurst, 2018; Meyer et al., 2018). The former involves people on prepayment meters not putting money into their meters and therefore being without access to energy in their domestic context. The latter refers to people deliberately limiting their energy use for reasons of income and affordability. This would include things like not using the heating system in a house or not using appliances for cooking and is already exemplified in the extracts discussed above. These forms of under-consumption are highlighted in the data as important foci for attempts to address energy poverty. However, in this section, I want to look in more detail at such forms of practice and examine the role that political discourse and policy plays in their normalisation as part of everyday life.

The research here highlights issues that go beyond ideas of self-disconnection and self-rationing—both of which infer something of a conscious somewhat calculated choice that is a temporary measure. Instead, I show how self-disconnection is often a regular and *enforced* part of coping with insufficient incomes that mean people cannot afford basic energy costs, and how self-rationing is for many normalised as part of everyday practice and connected to feelings about the self (such as worthlessness) that are, in part, engendered by policy. In these cases, the issues are not ones of a calculated short-term self-disconnection or rationing but concern 'living without energy use' as a normal part of life for reasons of unaffordability. Crucially, this is not necessarily related to the effectiveness of a heating system or building fabric, nor to access to suitable infrastructure or appliances, for example in transport or cooking. Rather, it concerns daily living practices in which energy services are to some extent considered a luxury good that a person can be expected to live without. With this first extract, I highlight the contention that self-disconnection is often not a conscious or calculated decision but something that people are forced to live with:

> ...every morning I wake up freezing cold. I know I could put the heating on, but say I put it on... like it doesn't maintain where there are so main draughts... You just learn to live with it. I think that's what you do... It makes you feel worthless. You haven't got no place in the world because nobody even knows or cares you're even here, because the government's just paying for you. (Biographical Interviewee 6, Bristol)

This extract attests to the ongoing and persistent nature of experiences of living without warmth. Notions of self-disconnection or self-rationing appear wholly inadequate to characterise the experience of waking to the cold every day. Importantly, in this quote, the pejorative welfare narrative of undeserving subjects, rather than the more positive discourse of fuel poverty, is reflected in the description of being made to feel 'worthless'. This suggests the overwhelming dominance of prevailing narratives about wider poverty, such that any subversions that might be offered by 'fuel' poverty as a specific category do not readily connect with or shape lived experiences and affective engagement. Instead, lack of access to energy services feeds into the negative individualised narratives of undeservedness. This is particularly problematic if considered in terms of the ways it may limit self-identification of energy poverty and thus be detrimental to responses and wider political mobilisation (cf. Petrova, 2018). In this next quote, as elsewhere in the research, the focus is on the normalisation of self-rationing as part of everyday life.

> I don't put the heating on until it is freezing, I've got blankets and throws everywhere, I'll just put a jumper on and put that over my knee, I don't put the heating on until it is really cold because it costs a lot of money and I don't ... It does get warm, if you have that one on and the little one on but it just costs so much money, I'd rather just put a jumper on, *I'm used to living in the cold, when I was young we didn't have any heating, we just had a fire when I was little so I'm used to cold, it doesn't bother me! It's fine.* (Biographical Interviewee 1, York)

In this extract, the participant—who is in receipt of disability benefits—discusses not using their heating unless 'it is really cold' as a normal part of her life and characterises it as 'fine' explaining how the cold 'doesn't bother' her. This can be related to other work on energy vulnerability and precarity where participants often reject the characterisation of themselves as living in fuel poverty (Day & Hitchings, 2011; Petrova, 2018). But such narratives can also be seen as related to the normalisation of energy precarity as part of everyday life (Petrova, 2018). These forms of positioning are potentially exacerbated by decarbonisation discourses that stress self-restraint and rationing without recognition of those people who should be using more to maintain basic capabilities, such as health and wellbeing. This also resonates with a concern that runs through the capabilities literature relating to how to determine or understand needs in any given context (Day et al. 2016). The normalisation of deprivation is important in this respect as it speaks to challenges in characterising needs from bottom-up perspectives.

The forms of accepted energy deprivation and under-use at issue here were common across the participants' narratives, with many not identifying non-use of energy as an issue but accepting it as a normal part of life. Again, in this next extract, the participant—who in this case has young children—discusses not using the heating as a way of coping with low income and unaffordability of energy.

> I don't really use the electric fire, it's more for show... so [I cope] by not using things really. The heating only goes on if it's really cold, things like that. (Biographical Interviewee 5, York)

Such descriptions of lives characterised by energy self-rationing in ways that affected capabilities extended far beyond heat to other forms of energy use, both within and beyond the home. For example, participants discussed processes of 'cutting back' on many other energy services from transport to communication technology.

> *Interviewer*: "Could you tell me a bit more about the impact of that [benefits cuts]?"
>
> *Interviewee*: "You know, having to cut back on shopping, gas and electric, having to cut back on going places in the car... So it's sort of luxuries, little luxury things... It's like my mobile phone..." (Biographical Interviewee 7, Bristol)

Here, the participant characterises multiple basic energy services as 'luxuries' further exemplifying normalisation of expectations surrounding lack of access to energy services. In addition to domestic uses of gas and electricity, this participant refers to energy services associated with transport and information and communication technologies. These energy services have been highlighted in research as having heightened significance for multiple capabilities within contemporary life (Day et al., 2016; Mattioli et al., 2017; Petrova, 2018; Simcock et al., 2016). Indeed, several participants cited transport needs, in particular, as something they would often forego out of necessity, again often normalising this as part of everyday practices.

> The only thing that I think that does impact us going out sometimes is bus fare and things like that. Really since we've moved... I feel really bad because we have not left [the area], we're always stuck here. I know it sounds silly, but £4 for a bus can be quite expensive sometimes.... ...We walk everywhere really, yeah. ... (Biographical Interviewee 3, Bristol)

Within the literature on transport poverty, the focus has tended to be on affordability of fuel and tendencies for lower-income groups to both live in areas with fewer transport options and drive older less efficient cars or be subject to enforced car ownership (Mattioli, 2017; Mattioli et al., 2017). As in the example of heating above, there has been little examination of people that do not access transport at all, instead opting to walk—sometimes very long distances—as their only option. Importantly, this set of quotes attest to the ways that these forms of deprivation, which relate to capabilities in multiple ways, are normalised. Such processes of normalisation by those subject to the welfare system connect back to literature that highlights how oppression and marginalisation shape what people see as acceptable in terms of human needs (Deneulin & McGregor, 2010; Mahali et al., 2018). Crucially, I argue here that discourses of welfare and undeserving subjects appear, in part, to shape expectations regarding access to energy services in ways indicative of an acceptance of energy poverty. This is revealing for the ways that poverty and energy poverty overlap in terms of the differing discursive narratives that characterise these policy areas (Middlemiss, 2016). Though as highlighted above fuel poverty is characterised by less pejorative discourses than welfare policy, the analysis highlights now this does not necessarily translate into experiences.

The research data in this chapter speak to the foundational nature of energy poverty in compounding and perpetuating poverty more widely. The challenges of being able to access employment, for example, are highlighted and the role of energy services in the cyclical nature of wider poverty is brought to the fore (Macdonald et al., 2020; Mattioli et al., 2017). The limited direct support for energy needs within welfare policy arguably also compounds processes of normalisation around lack of access to energy services, as people are met with expectations for them to be able to fulfil such needs, such as travel to interviews, without (further) support—at least not at the point it is required. These normalised experiences of living without energy use, of not using transport methods other than walking, of not turning heating on, or of not cooking or using lighting, also create issues in identifying people that need help (including in cases where local services make referrals). The below quote from one of our local stakeholder participants highlights this issue.

> I suppose the other bit for me is I can sit here and say I work with very vulnerable people, but actually they might not think themselves as vulnerable. When you're talking about fuel poverty…we're talking about *generations that have had nothing, so actually that's normal for them, that's their ordinary life*…. there are a huge amount of people that won't [take help] and they will sit in their living room and they will go cold and they won't eat meals, or they'll just provide for their family, because they don't see themselves as vulnerable, they're just surviving. (Stakeholder Interviewee 3, York—Workshop)

This is again indicative of the normalisation of energy poverty and its deep relation to entrenching poverty more generally. Such processes of normalisation and institutionalisation are damaging, then, in perpetuating poverty precisely because they affect the extent to which people do (or even can) mobilise to change their situations. Petrova's (2018) research has highlighted normalisation and institutionalisation of energy deprivation specifically amongst young people living in Houses in Multiple Occupation (HMOs) and the private rented sector. She asserts that the production of precarity reflects governance processes and approaches that enshrine home ownership as 'proper' housing consumption and neglect strategic interventions that ensure decent housing provision in these sectors.

The data discussed here chime with Petrova's analysis but suggest further that acceptance of energy deprivation is pervasive, reaching across different working-age cohorts affecting people living in very different circumstances, and even those typically positioned as vulnerable within policy (e.g. disabled people, young children). In the case of this research, focusing as it does on people affected by welfare policy, the mechanisms of 'bearable acceptance' identified by Petrova (2018, p. 26) are comparable with the troubling forms of complete normalisation that here can be seen as arising, at least in part, from politicised individualised discourses of 'unworthy' and 'undeserving' subjects. I argue such normalisation also arises from the constitution of expectations for accessing energy services and engaging in forms of practice, without recognition of the constraints people face. But this is addressed further in the following chapter.

In Petrova's research, the temporality of the present was a salient aspect of precarious situations as the young people that were the focus of her study accepted poor living conditions based on their being provisional and non-permanent (even if this in some cases was not borne out over time). There is a temporality too in the narratives of energy deprivation arising from the research here, but as the quotes above attest this was one of cyclicity and perpetual precarity (see also Macdonald et al., 2020), such that this becomes accepted as a normal feature of life. In this instance, then, it is the ongoing—rather than short term—nature of the difficult conditions in which people are living, allied with entrenched narratives of undeservedness, that serve to institutionalise energy poverty and decrease the space for political contestation. Arguably, this space is also further limited by the separation between fuel poverty and wider poverty, which though important and 'pioneering' (Bouzarovski, 2018, p. 10) in many respects belies a lack of interrogation of the connections between them, such as the central significance of energy services in contributing to the cyclical temporalities that characterise poverty.

Concluding Discussion

By looking at an area of invisible energy policy—namely welfare policy—and examining the ways that it shapes capabilities related to energy services, the analysis in this chapter has taken a first step toward realising some of the insights afforded by drawing together these areas of energy research. In this conclusion, I reflect on the implications of the analysis for advancing understanding of energy poverty, including what is gained by shifting the analytic focus beyond energy policy.

To begin, it is clear from the analysis presented here that the by now widely accepted argument of fuel poverty extending well beyond contemporary preoccupations with heat and older people is crucial for understanding problems of energy deprivation. The focus in this research was on working-age people, intentionally excluding analysis of welfare policies related to older people as this group has been given the most attention historically in energy poverty analysis and policy (Simcock et al., 2016). Examining working-age people has been revealing both for thinking about other groups (beyond older people) and for considering the ways that energy is intricately interwoven with multiple capabilities essential for social participation (e.g. relating to work), including but also going far beyond heat and domestic settings. This chapter thus speaks to the value of an expanded understanding of energy poverty as affecting multiple different groups and entailing access to energy services to support capabilities, but takes this further to suggest the value of such an approach for more fluid engagement with multiple domestic energy uses *and* mobilities.

By looking at welfare policy as an area of invisible energy policy, this research has brought into view the ways that policies outside of energy have important implications for energy services that connect with people's capabilities, such as those related to health and wellbeing, but also income and social respect. This brings focus onto invisible energy policy as an important agenda for understanding not only the formulation of increasing patterns of energy consumption and over consumption, but the reproduction of under-use and energy poverty. In this, I highlight both more direct forms of policy intervention and wider political narratives and framings as important foci for analysis (see Butler et al., 2018). Welfare reforms have been examined here for their more direct roles in shaping people's energy-related practices and experiences of under-use, affecting basic capabilities. But the wider discourses that pervade welfare policy and fuel poverty policy domains have also been examined for their role in shaping processes of normalisation and the social reproduction of energy poverty. Overall, the analysis of welfare policy has been important in bringing into view the relations between energy poverty and wider poverty, highlighting the importance of attentiveness to the interconnections.

Much prior research has tended to focus on emphasising the distinctiveness of energy poverty given the hard-fought battles to have it recognised as a separate issue with requirements for policy to address it. In this context, while there are clear links between welfare policy and energy deprivation, and between fuel poverty and wider poverty, it is not as simple to conclude that poverty should be the focus of policy, rather than fuel poverty. There is history here and there are positives to the identification of fuel poverty as a distinctive issue—not least the budgetary allocations and obligations on suppliers to fund supportive measures to address this issue (Bouzarovski, 2018), as well as the depoliticisation of the subjects of fuel poverty policy in contrast to welfare policy (Middlemiss, 2016). However, there are issues identifiable in the research here concerning how such positives translate into lived experiences.

Centrally, this research offers insight into how the pejorative discourses of welfare policy appear as far more dominant and prevailing within the narratives of many of those experiencing (energy) poverty, than the 'worthy' subjects of fuel poverty policy. Indeed, not only here but across the growing literature on lived experiences of fuel poverty more generally, it is difficult to find a case where such depoliticised understandings of what it is to be subject to fuel poverty are expressed (e.g. Chard & Walker, 2016; Willand & Horne, 2018). This means that while it is useful within policy and politics, it is does not necessarily have such effects within the lived experiences of energy poverty. One consequence arising, at least in part, from the inculcation of such subjectivities is the apparent normalisation of severe energy deprivation (Petrova, 2018). Such normalisation is problematic in terms of the way it constrains political mobilisations around the issues, preventing energy poverty from being brought to light. And it is relevant to challenges of identifying and targeting help and support as people do not self-identify as in need of help or recognise that they would be entitled. Though there is significant scope within policy for local implementation and allocation of fuel poverty measures, which can better attune to specific circumstances, this does not fully overcome challenges in contexts where entitlement is both derided and obscured.

This is not a call, then, for the subsummation of fuel poverty into the wider category of poverty, but rather for the recognition of energy poverty as an even more fundamental underpinning to multiple social issues. Crucially, the research here reveals that energy deprivation contributes in fundamental ways to the cyclical patterns of poverty identifiable within wider literature (e.g. see Macdonald et al., 2020). Highlighted here are the ways that different forms of energy poverty, across domestic lives and within mobilities, relate to capabilities in fundamental ways. Where other societal problems, such as health, are often in focus when it comes to energy poverty, the tendency has been to avoid confrontation of the relations between energy poverty and wider poverty. However, I argue that the importance of addressing energy poverty is underscored, rather than undermined, precisely by recognising its connections to wider poverty and the ways these relations are important to processes of reproduction.

Finally, the analysis also nods toward the importance of thinking about the politics of under-use, its normalisation and social reproduction, together with practice theory-based ideas about the specification of need (Shove et al. 2012). In the next chapter, I turn to wider concepts from practice-theory based analyses to examine the dynamics relations through which need for energy is constituted. This takes in the wider systemic and seemingly non-energy-related processes contributing to creating needs and examines them as central to understanding the reproduction of energy poverty. The analysis brings recognition of the ways that the patterns and trends that shape rising consumption for those with financial means, also have implications for the capabilities of those without. These arguments are advanced by the analysis in the next chapter, where I move to focus on the constitution of energy demand and the ways that this, which is typically the focus of practice-based energy sustainability research, can be revealing for understanding energy poverty.

References

BEIS (Department for Business, Energy and Industrial Strategy). (2021). *Sustainable warmth: protecting vulnerable households in England* [online]. https://www.gov.uk/government/publications/sustainable-warmth-protecting-vulnerable-households-in-england

Bouzarovski, S. (2018). *Energy poverty: (Dis)assembling Europe's infrastructural divide*. Palgrave.

Butler, C., Parkhill, K. A., & Luzecka, P. (2018). Rethinking energy demand governance: Exploring impact beyond "energy" policy. *Energy Research & Social Science, 36*, 70–78.

Chard, R., & Walker, G. (2016). Living with fuel poverty in older age: Coping strategies and their problematic implications. *Energy Research & Social Science, 18*, 62–70.

Cox, E., Royston, S., & Selby, J. (2019). From exports to exercise: How non-energy policies affect energy systems. *Energy Research & Social Science, 55*, 179–188.

Day, R., & Hitchings, R. (2011). "Only old ladies would do that": Age stigma and older people's strategies for dealing with winter cold. *Health & Place, 17*(4), 885–894.

Day, R., Walker, G., & Simcock, N. (2016). Conceptualising energy use and energy poverty using a capabilities framework. *Energy Policy, 93*, 255–264.

Duffy, S. (2014). *Counting the cuts: What the Government doesn't want the public to know*. Centre for Welfare Reform on behalf of the Campaign for a Fairer Society.

Deneulin, S., & McGregor, J. A. (2010). The capability approach and the politics of a social conception of wellbeing. *European Journal of Social Theory, 13*(4), 501–519.

Großmann, K., Buchholz, J., Buchmann, C., Hedke, C., Hoehnke, C., & Schwarz, N. (2014). Energy costs, residential mobility, and segregation in a shrinking city. *Open House International, 39*, 14–24.

Großmann, K., & Kahlheber, A. (2017). Energy poverty in an intersectional perspective: On multiple deprivation, discriminatory systems, and the effects of policies. In N. Simcock, H. Thomson, S. Petrova, & S. Bouzarovski, *Energy poverty and vulnerability: A global perspective* (pp. 12–32). Routledge.

Hargreaves, T., & Longhurst, N. (2018). *Consultation Response: Ofgem Prepayment self-disconnection and self-rationing—A call for evidence* [online]. https://www.ofgem.gov.uk/sites/default/files/docs/2019/08/centre_for_competition_policy_response.pdf.

MacDonald, R., Shildrick, T., & Furlong, A. (2020). "Cycles of disadvantage" revisited: Young people, families and poverty across generations. *Journal of Youth Studies, 23*(1), 12–27.

Mahali, A., Lynch, I., Wilson-Fadiji, A., Tolla, T., Khumalo, S., & Naicker, S. (2018). Networks of well-being in the global South: A critical review of current scholarship. *Journal of Developing Societies, 34*(4), 373–400.

Mattioli, G. (2017). "Forced car ownership" in the UK and Germany: Socio-spatial patterns and potential economic stress impacts. *Social Inclusion, 5*(4), 147.

Mattioli, G., Lucas, K., & Marsden, G. (2017). Transport poverty and fuel poverty in the UK: From analogy to comparison. *Transport Policy, 59*, 93–105.

Meyer, S., Laurence, H., Bart, D., Middlemiss, L., & Maréchal, K. (2018). Capturing the multifaceted nature of energy poverty: Lessons from Belgium. *Energy Research & Social Science, 40*, 273–283.

Middlemiss, L. (2016). A critical analysis of the new politics of fuel poverty in England. *Critical Social Policy, 37*(3), 425–443.

Middlemiss, L., Ambrosio-Albalá, P., Emmel, N., Gillard, R., Gilbertson, J., Hargreaves, T., Mullen, C., Ryan, T., Snell, C., & Tod, A. (2019). Energy poverty and social relations: A capabilities approach. *Energy Research & Social Science, 55*, 227–235.

Morris, R. (2013). 'Unjust, inhumane and highly inaccurate': The impact of changes to disability benefits and services: Social media as a tool in research and activism. *Disability & Society, 28*(5), 724–728.

Mould, R., & Baker, K. J. (2017). Documenting fuel poverty from the householders' perspective. *Energy Research & Social Science, 31*, 21–31.

Pemberton, S., Fahmy, E., Sutton, E., & Bell, K. (2015). Navigating the stigmatised identities of poverty in austere times: Resisting and responding to narratives of personal failure. *Critical Social Policy, 36*(1), 21–37.

Petrova, S. (2018). Encountering energy precarity: Geographies of fuel poverty among young adults in the UK. *Transactions of the Institute of British Geographers, 43*(1), 17–30.

Roulstone, A. (2015). Personal Independence Payments, welfare reform and the shrinking disability category. *Disability & Society, 30*(5), 673–768.

Royston, S., Selby, J., & Shove, E. (2018). Invisible energy policies: A new agenda for energy demand reduction. *Energy Policy, 123*, 127–135.

Shove, E., Pantzar, M., & Watson, M. (2012). *The dynamics of social practice: Everyday life and how it changes*. Sage.

Simcock, N., Walker, G., & Day, R. (2016). Fuel poverty in the UK: beyond heating? *People Place and Policy Online, 10*(1), 25–41.

Willand, N., & Horne, R. (2018). "They are grinding us into the ground"—The lived experience of (in)energy justice amongst low-income older households. *Applied Energy, 226*, 61–70.

CHAPTER 6

Energy Poverty, Practice, and Inequality

Abstract This chapter builds from the previous one to develop a focused analysis that explores the value of key ideas from practice theory-based energy research for analyses of energy poverty. The chapter uses examples from in-depth qualitative research to give attention to the *constitution of need* in terms of implications for energy deprivation, as well as examining the ways that power relations shape processes of recruitment and defection from practices. In this, it offers distinctive research trajectories for energy poverty research by extending beyond concern with people's abilities to *meet* needs or achieve capabilities. And it offers a basis for response to critiques of practice-informed analyses of energy demand that highlight the limited attention given to inequalities within such work.

Keywords Energy poverty · Practice theory · Inequality · Welfare policy · Capabilities

Introduction

The previous empirical chapter drew on capabilities thinking to understand energy poverty and developed an analysis that explored how welfare policy—as an area of invisible energy policy—shapes experiences of energy deprivation. This current chapter retains the capabilities-based understanding and exploration of the invisible energy policy area of welfare

C. Butler, *Energy Poverty, Practice, and Policy*, Progressive Energy Policy,
https://doi.org/10.1007/978-3-030-99432-7_6

but brings focus onto the constitution of energy demand and the reproduction of social practice. Building on the previous chapter's analysis and formulated as an empirically informed discussion, the focus here is on how a concern with the constitution of need, on the one hand, and processes of enrolment in practices, on the other, can offer distinctive trajectories for energy poverty research. And conversely, how in working through such questions, it becomes possible to build insights important too for practice theory-based work on energy demand, centrally around the relevance of attentiveness to relations of power for analyses.

One of the key interventions of practice theory-based energy research has been to argue for focus on the ways that energy demand is constituted and specified (Shove, 2003) (see Chapter 2 for discussion). Much energy demand policy has tended to address ways of meeting existing levels of service using less energy (e.g. through efficiency or behavioural interventions), or changing technologies to reduce the environmental damage arising from energy use (e.g. renewable energy and electric cars). Shove (2003, p. 396), however, has argued for a focus on more 'penetrating questions' that concern the processes through which services are specified and constituted in the first place. In a global context where the energy intensity of daily life is ever increasing, she asserts that the core question should be: 'How do new conventions become normal, and with what consequence for sustainability?' (ibid.). Though this question has been of central significance for work on energy and sustainability, I argue here that it raises equally important questions for energy poverty research.

The previous chapter engaged with debates at the forefront of energy poverty that have brought focus on how energy is foundational for multiple capabilities. These debates have dramatically opened up the focus of research and practice to a wide range of energy services beyond the more conventional emphasis on heat (Bouzarovski & Petrova, 2015; Day et al., 2016; Middlemiss et al., 2019; Petrova, 2018; Simcock et al., 2016). Research in this space has also advanced to address the complex and multifaceted processes that shape experiences of energy deprivation, highlighting wider structural conditions and processes at play. However, the emphasis remains on the factors and processes that affect abilities to meet energy service needs, without delving more deeply into the relevance of ideas about how those needs are created and constituted.

Research focuses on the ways that people's abilities to meet energy service needs are reduced or affected by social, political, and economic processes, as well as through personal circumstances, and structurally constituted material conditions (Bouzarvoski & Petrova, 2015; Day et al., 2016; Middlemiss & Gillard, 2015; Petrova, 2018). And research has examined the relations between energy deprivation and people's capabilities, showing the complex interconnections and deleterious effects (Middlemiss et al., 2019; Mould & Baker, 2017). Very little attention has been given, however, particularly in empirical research, to the processes through which energy service needs are *constituted* with consequences for energy poverty. While Day et al. (2016, p. 262) comment on the ways that the capabilities approach (see Chapter 2 for discussion) 'allows us to see the effect of evolving social norms in constituting energy demand and, therefore, relative energy deprivation', this has yet to be taken forward in any significant way within analyses.

At the same time, as discussed in Chapter 3, practice theory-based research has been focused on the constitution of energy demand but has tended to address the consequences for sustainability giving limited attention to poverty and inequality. Indeed, with two notable exceptions focused on conceptual contributions, namely a 2002 working paper by Shove and a 2013 book chapter by Walker, there has been very little reflection on inequality in energy research inspired by practice theory at all (see Chapter 3 for a detailed discussion). The turn to invisible energy policy within practice theory-based work offers some important avenues for bringing focus on relations of power as it opens up to questions about the processes through which practices are shifted and shaped through governance. Through the discussion here I show how a focus on welfare policy gives a view of relations of power not only in the production of conditions of energy poverty but in the constitution of needs for energy and in abilities to be recruited to, or defect from, practices.

In the following discussion, I thus develop an analysis that raises and addresses several key concerns that emerge from bringing practice theory thinking (including that relating to invisible energy policy) to bear on issues of energy poverty and capabilities. The first relates to the specification of need and how practices-as-entities come to be constituted in ways that exclude people and create energy deprivation, emphasising the active role of policy in these processes. The second connects more directly with Walker (2013) to highlight how people have varying degrees of agency (as well as capability) as performers of practices and are thus differently placed to be recruited to or defect from practices. A key argument that is developed here concerns how this is related to the ways that practices are shaped or steered across distinct policy spaces through particular formulations of power relations, some of which are more coercive than others. I argue that in the case of welfare policy, possibilities for defection from practices are severely constrained regardless of capability to engage in those practices. This analysis thus weaves a central concern with the role of governance and policy in constituting need and shaping practice into the interventions made within the previous empirical chapter. By working with the empirical data and combining conceptual traditions utilised across energy demand research, the analysis presents a unique contribution to debates about energy poverty, practice, and policy.

Exploring Power in the Constitution of Need

Where the previous chapter utilised examples from across the spectrum of energy services, this chapter focuses in further on two key examples—those of information and communication technologies (ICTs), and travel and mobilities. These examples are utilised to structure the discussion here because they speak most readily to the ideas this chapter explores relating to the constitution of needs and processes of recruitment and defection from practices. This is partly because they are identifiable as interrelated areas where welfare policy reforms have been important in constituting newly emerging and increasing needs. As discussed in the previous chapters, though such energy services are not typically the focus of energy poverty research, both have important relevance from an energy capabilities perspective. ICTs have been highlighted as an increasingly important energy service particularly for different groups outside of older people (see Petrova, 2018; Simcock et al., 2016). And although concern with mobilities has long been foreshadowed in the cognate but distinctive transport poverty literature, it has yet to be brought into analyses of energy capabilities in a prominent way (see Mattioli et al., 2017; Robinson & Mattioli, 2020). This makes these areas of energy service of additional interest and importance in many respects, providing further context and grounding for the analysis that follows.

Constituting Digital Worlds

The first examples I discuss relate to needs for information and communication technologies (ICTs) and people's inability to access these energy services with implications for capabilities. While ICTs have not been given a great deal of focus in energy poverty research or energy demand research more widely, they are receiving increasing attention as understanding of their energy implications grows (Morley et al., 2018) and their importance to capabilities becomes ever greater (Simcock et al., 2016). Simcock et al. (2016) refer to new consumer electronics becoming 'basic necessities' for life in the UK. They emphasise how ICTs as a set of energy services are increasingly integral to daily life and capabilities underpinning wellbeing (Petrova, 2018; Simcock et al., 2016; Walker et al., 2016). Recent global events (including the COVID-19 pandemic) have seen even greater acceleration of trends towards ICT use adding to its importance in daily life and its relevance as a growing area of home electricity use (Morley et al., 2018). Even if digital services are not considered costly at point of energy use, such services require *continuity* of electricity supply for regularly charging devices, which becomes far more problematic in the context of self-disconnection, under-use, and enforced (if temporary) disconnections associated with prepayment meters.

These services are also bound up with requirements for hardware and data contracts that are more expensive and burdensome. Understanding energy deprivation in terms of energy services and the capabilities that they relate to, then, highlights not only the point of use energy expenditure but the equipment, service contracts, and other costs associated with the relevant services, such as computers, mobile phones, and WiFi connections (Bouzarovski & Petrova, 2015). This might be thought of as entirely out of kilter with current policy definitions of fuel poverty, but it is possible to argue that it simply addresses a similar aspect of service to that of boiler installation, grid connection, and electrical points and light fittings to which policy is already directly targeted. That is to say, the materials and services related to ICTs are arguably equally central to enabling energy services as these wider core forms of energy infrastructure.

Moreover, in highlighting how ICTs have come to be prioritised, Petrova (2018) shows that they increasingly figure in processes through which different needs are balanced, affecting under-use of other services such as heat. This both points to the growing importance of ICTs for contemporary life and shows how areas of new and/or increasing need interact with other energy services. There are good reasons, then, for considering them to be significant in terms of energy poverty concerns. The analysis that follows contributes to further understanding the relevance of ICTs to issues of energy poverty. However, the central focus is on using this example to cast light on the role of (invisible energy) policy in *constituting needs* (cf. Morley et al., 2018) and, crucially, on the ways that people have differential abilities for resisting recruitment to, or defecting from, practices with important implications for experiences of energy poverty.

Amid wider policies of digitalisation (Morley et al., 2018), welfare policy has seen increasing emphasis on digitalisation with requirements for anyone in receipt of benefits to access everything online (Butler et al., 2018). This includes applications for benefits and any changes to circumstances, job applications, and other tasks related to finding work (which those in receipt of certain benefits are expected to complete as part of Claimant Commitments—Citizens Advice, 2020), as well as information about processes and procedures to which they must adhere. This has increasingly made ICTs essential as reflected in the accounts of the research participants: '*...everything's now over the phone isn't it, computers... I'll wait until I'm on a good day and do it then*' (Biographical Interviewee 7, York). While there are many areas of life that are difficult to negotiate without access to ICTs meaning that enrolment in digital worlds is a reality for most in the UK, I argue that the nature of 'recruitment' for those subject to welfare systems is deeper still. The welfare system has very explicitly and actively been digitalised through government policy, with an arm of the Department for Work and Pensions—DWP Digital—dedicated to overseeing and advancing these processes:

> We have recently launched over a dozen digital services, including the Universal Credit, Carers, and Pensions services... which improve outcomes for 22 million people. (DWP Digital, 2016)

These processes are cast in terms of improvement but in the context of welfare reform, they have also been imposed on claimants with requirements to make and manage claims digitally; first by phone but over time this has shifted to online:

> Instead of a 40-minute telephone call involving both a customer and an agent, online applications now take on average 18 minutes. (DWP Digital, 2021)

The language of efficiency and improvement, which colours much of the discourse around these processes, elides how expansion of digital has happened while other ways of accessing services have been reduced or restricted. Since the early 2000s, jobcentres and places for in-person access have gradually been reduced and moved to central locations in cities. This is particularly restrictive as, in the UK, city centres are expensive places to live and most of those using the welfare system as a primary source of income are therefore unlikely to live close to service locations.

> ...services in York, everything was taken out of the community and put into one central place which is in the city centre which is great for saving money but if you've got a family that live out on the outskirts... and they need to get into York, well it's not easy to do that by bus. (Stakeholder Interviewee 10, York)

Processes of closing jobcentres and digitalisation have accelerated more recently with over 100 further closures of jobcentres between 2016 and 2018 (Finn, 2018). Such is the extent of closures that the aftermath of the global coronavirus pandemic has seen temporary jobcentres opened as an emergency measure to cope with the increased demand for welfare services. These processes of closure severely limit other ways of using welfare services reinforcing requirements for digital access.

> *All their benefits they've got to apply online, jobs are all now online, everything's online.* So a working phone and a good one is becoming absolutely essential, almost as essential as food. And the poorer you are, the more essential it is. (Stakeholder Interviewee 13, Bristol)

Digital services are thus actively instituted through policy and governance processes as part of the requirements for engaging with welfare services. These governance processes are introduced in ways that highlight positives for access without consideration of the implications for energy services and needs. In this way, they can be considered part of the invisible energy policies that characterise wider shifts to digitalisation (see Morley et al., 2018). However, the above quote also highlights important issues in processes of 'recruitment' to digital practices for those implicated in the welfare system. Though by taking up and engaging with ICTs as part of daily life, people are active in constituting the related practices, there are differences evident here in the extent to which people can defect from such processes of enrolment. In the case of welfare policy, the requirements for digital access that are generative of new energy service needs involve recruitment to practices with very limited opportunities for defection. It has become an essential means for accessing their only source of income.

Where the quote above highlights how 'the poorer you are, the more essential it [access to ICTs] is', crucial to this story is how such processes of constituting need combine with a lack of access to these energy services. The research highlights how people are often unable to meet these needs with knock-on consequences for multiple capabilities relating to securing their income, as well as other dimensions of wellbeing, such as social respect, social relations, and mental health. Many of the participants in the study discussed challenges that arise in gaining access to information and communication technologies required for engagement with the welfare system. This included challenges in securing or negotiating access, problems with continuity in electricity for charging devices related to prepayment meter use, and lack of wider service provision outside of the home.

Interviewer: So you haven't got a computer or the internet at home?
Interviewee: No I have to go to the job centre to check on their computers and if it's out of hours then I do try and ask a neighbour but they get a bit funny with me asking... (Biographical Interviewee 8, Bristol)

While the energy service needs related to ICTs are something that could be met communally, these increasing requirements are being constituted at the same time as access outside of individual homes has been eroded, particularly for those enrolled in the benefits system. As highlighted above, jobcentres have closed and been relocated but so too have other services that offer points of free access, such as libraries (Butler et al., 2018; Finn, 2018). This heightens requirements for in-home forms of access, then, that push costs and energy service needs onto individuals and households. The below quote from an interview with a couple in receipt of disability benefits articulates problems related to the limited access to ICTs people in the welfare system have.

Interviewee 6b: There's not enough computers at the Job Centre and the council office combined because there's so many people who cannot afford to have the internet connection, but then they kick you off them after a certain time anyway... The vast majority of their customers... don't have any internet at all ...
Interviewee 6a: And don't have a device to connect to the internet either.

Interviewee 6b: Some piggyback off neighbours... (Biographical Interviewees 6a & 6b, York)

In this context, people are subject to specific requirements for this energy service without the requisite abilities to meet needs. Crucially, these requirements are not simply part of wider processes that are benignly unfolding, they are actively constituted and created through and within welfare policy with detrimental implications for capabilities. In terms of energy poverty research, this highlights the importance of giving focus to the processes through which energy service needs are constituted. Such a focus within analysis could bring to the fore multiple issues related to the ways energy poverty is understood and addressed. For example, the income thresholds that form part of current policy do not readily recognise the emergence of 'new needs' (Walker et al., 2016) or, more fundamentally, this could advance recognition of how reducing needs for energy can be as important for addressing energy poverty as it is for problems of environmental sustainability.

The role of policies (beyond energy policy) in constituting needs is frequently obscured in both policy and analysis, as are the ways that different vulnerabilities or forms of precarity are affected by such changes. This chimes with existing research that has argued for greater focus on intersectionality and the wider systemic structures and inequalities that underpin energy deprivation (Großmann & Kahlheber, 2017; Middlemiss & Gillard, 2015; Petrova, 2018), but goes further in bringing to the fore the relevance of the constitution of need within and through policy to the (re)production energy poverty. Importantly, the analyis here also gives focus to that ways that people have differential abilities to navigate, negotiate, or resist emerging trends and the policy interventions that underpin them.

Walker (2013) has suggested that in thinking about inequality, analysis might look at the ways that people are 'unrecruitable' to practices owing to their not having the requisite capabilities. For example, he suggests that a person may be 'unrecruitable' if they lack capabilities related to the physical, material, or skills elements required to engage in a practice. In the case here, however, people can be seen as both 'unrecruitable' as Walker (2013) has proposed, in lacking capabilities to engage in practices related to ICTs, and also subject to a form of *enforced recruitment* to those practices. This is indicative of limited agency in the potential people have for defection or resistance in the processes through which practices become enshrined.

Practice perspectives have tended to emphasise processes through which practice-as-entities become established or diminish when cohorts of people take them up or abandon them. Bringing inequality more firmly into a practice theory perspective highlights how as new practices take hold, and are enacted by multiple people, some are increasingly marginalised owing to limitations in their capacities to engage in such practices. Moreover, evident here is that people are both differently placed to *enact practices* that have taken hold in the wider population and to *resist enrolment* in processes of uptake as some are forced to engage by virtue of their socio-economic position and the policy spheres to which they are subject. In Gormally et al.'s (2019) paper, they discuss the ways that education policy is actively negotiated in daily life and to some extent resisted by those working in higher education. Noticeable from the research here is that welfare policy exerts a far more coercive force upon its subjects, highlighting the differential ways that policy acts upon citizens. This adds a further layer to understanding of the processes by which practices take hold and brings inequality into view, not only in terms of how it affects abilities to engage in practices but how it shapes possibilities to subvert existing trends and/or constitute new ones.

Constituting Mobilities

The second example I want to address in working through ideas about the constitution of need, recruitment, and defection relates to mobilities, travel, and transport. As discussed in Chapter 2, though energy poverty research has tended to be dominated by a focus on domestic energy use, there is a body of research focused on transport poverty (e.g. see Mattioli, 2017; Mattioli et al., 2017 for discussion). This research has developed important insights relevant to understanding processes of energy vulnerability but has not yet engaged with the capabilities-based thinking that informs key parts of the domestic energy poverty literature (Bouzarovski & Petrova, 2015; Day et al., 2016; Middlemiss et al., 2019; Simcock et al., 2016). With focus on the implications of (non-energy) policy for mobilities and energy poverty, this section builds on the analysis above and in the previous chapter to bring further attention to the ways that needs have been constituted though policy and governance. This goes beyond the existing transport poverty literature to draw out questions about the ways that the creation of needs related to mobilities underpins experiences of energy poverty and relates to capabilities.

There are multiple interlinked ways in which welfare policies and reform can be seen to have created needs for travel. These concern, for example, policies discussed in the previous chapter such as requirements for claimants of unemployment benefits to demonstrate that they are actively seeking work. Part of this activity involves attending job interviews and keeping appointments at job centres. As highlighted previously, failure to attend on time results in sanctions, meaning deductions or even entire withdrawal of welfare benefits payments for a specified period. In the research, participants discussed these requirements for travel as often involving long distances with associated costs (e.g. for bus fare) being challenging to meet. While some travel costs can be reclaimed, often participants did not have the money to support the upfront payment on the day.

> [For a job interview] I lived in [Place name] and I had to walk to [Place name]. In the car it's probably about, I think about 40 minutes maybe or something, and I walked it. So I was walking in the rain, so already you look a mess, and you haven't got credit to ring them, so they think you're not interested, and because you can't afford the bus fare there, it's just you're constantly going round in circles. It's really difficult and mentally it's very hard to keep pushing forward, and you can't go to the Job Centre and be like oh, I can't get to my interview this day.... (Biographical Interviewee 6, Bristol)

The requirements to attend appointments and interviews highlighted here are one way in which welfare policy is constitutive of requirements for travel, but there are other policies that affect the places where people live and the locations of service infrastructure too. This means that the distances people must travel (as well as the necessity for travel at all) are also affected by specific welfare policy changes. These changes include things such as: limitations placed on choice of housing location within the allocation system; the bedroom tax (which has seen occupants forced to move home to be able to afford housing); and reforms to the ways housing benefit is delivered under universal credit that mean it goes directly to the recipient instead of the landlord, with implications for the willingness of some to accept those on welfare support (also see

Chapter 5). This combines with closures of job centres and centralisation of services (discussed above) bringing further implications for travel distances. In combination all this speaks to the ways that policy shapes both the location and availability of services, the places where people live, and the requirements for travel. In focus here, then, are not only the challenges people face in meeting needs but the ways that these needs (*and* the challenges) are constituted by intersections between different policies.

The previous chapter pointed to work on transport poverty, which has been attentive to how transport needs vary across different groups. For example, Mattioli (2017) discusses the factors that affect where people live, and highlights work on urban socio-spatial configurations that shows how lower-income groups tend to live in areas prone to higher car dependence. Though this recognises the inequalities that exist in needs for travel, the analysis here goes further speaking to the ways that policy is directly implicated in the underlying processes that shape the emergence of these trends and highlighting differential abilities to both constitute and negotiate such needs. As above, the policies shaping travel requirements for those subject to the welfare system are far more punitive and draconian than might be reflective of other policy areas. This analysis here, then, aims to highlight not only how policy constitutes needs but also how different policy areas are characterised by different power relations. While the language of 'steering' might be appropriate for analysis of some policy areas, others, like welfare, are characterised by something far closer to coercion.

As the next example highlights, these welfare policy reforms compound and intersect with wider areas of policy and commercial activities that further shape and create challenges related to mobilities and energy poverty.

> There are pockets in Bristol of poverty and in those areas there aren't decent shops, you can't buy food unless you travel out… Bus routes don't go through there in the same way because it doesn't pay. And bus routes [that] were being subsidised [have] been withdrawn as well. So now travelling around is now becoming a big issue… they are employment blackspots… you're desperately trying to find a job and you go for an interview, get a job in [Place name]. But then you've got to get there somehow every day. And I've tried to do that as well. It took me two hours to get there and two hours back again and £5 a day as well. So people don't realise, it's not just food to eat, heat your home, travelling

> around, everything is so much harder, becomes harder, if you're in one of these low income situations. (Stakeholder Interviewee 13, Bristol)

This attests to the complex intermingling of policies and processes that shape where people subject to welfare policy live and how their energy requirements and experiences of energy poverty are exacerbated and reproduced. This reflects arguments made in the previous chapter concerning the cyclical nature of the relations between energy poverty and wider poverty. I argue here, however, that such circumstances do not arise by chance or represent instances of individual misfortune but are fundamentally shaped by policies that are constitutive of both where people (can) live and the services and facilities available to them within their communities. This concerns the ways, then, that energy services are *actively constituted* as necessary to achieve a minimally decent standard of living, while simultaneously being made inaccessible. This speaks once again to the points raised above about the relative abilities people have for resisting, negotiating, and reducing their engagement in different practices. In the case of welfare policies, imperatives for travel are such that people must find a way to meet them even if outside of their means and capabilities. Such variation in abilities to decide and negotiate engagement in travel practices could be attended to as a way to bring inequality into analyses of the increasing energy intensity of mobilities more widely.

For the final point I return to issues discussed in Chapter 5 about the ways that welfare is problematised through individualised stigmatisation of its subjects (see also Middlemiss, 2016; Wright, 2016). Where the emphasis is on the individual and notions of personal deficits in skills, willingness, or ability (for example), the ways that policy itself can be constitutive of problems of mobility and energy poverty are likely to be obscured or pushed out of remit. Understanding the boundaries, problematisations, and framings of policies, then, is likely to be important in both questioning the processes that are generative of new needs and examining where the spaces for intervention may lay within this.

Concluding Discussion

The energy demand literature discusses the constitution of needs, and concepts of recruitment and defection from practices, with primary focus on the implications for environmental sustainability. The discussion here has sought to bring emphasis onto the importance of these ideas for examining energy poverty and, as part of this, develop insights relevant to areas of analysis that have been relatively neglected within existing work. The first area relates to the invisible energy policy literature and the point that in work addressing questions about the constitution of need, the specifics of policy have not been the focus (Royston et al., 2018). The second area concerns the contention that the relevance of inequality to the reproduction of practices has rarely been addressed (Walker, 2013). This chapter has sought to emphasise the ways that welfare policies, in particular, are constitutive of needs and in doing so, bring to light issues of inequality by emphasising how people's agency in negotiating the requirements for energy services can be highly restricted, as well as being entwined with a lack of possibilities to meet them.

In the examples discussed above, this is borne out through highlighting the ways that needs for energy services related to both information and communication technologies (ICTs) and mobilities have been constituted through welfare policy. At the same time, policies are also shown to shape the possibilities people have for meeting these energy service needs. For instance, by creating conditions whereby communal access to ICTs is limited or in shaping processes of housing provisioning in ways that mean people are more likely to live at distance from service centres. The examples here indicate how in contexts of welfare policy, the degrees of agency available to people both in determining the nature of needs and the possibilities for defecting from practices are extremely limited. They attest to a situation where people are recruited to practices but without the requisite abilities to fulfil them. This analysis suggests, then, an important variability in agency arising from inequality and our relations to policy as subjects of different policy fields. Turning to Bourdieu (1998), the analysis here captures something that speaks to his notion of symbolic violence, wherein the ways that shared meanings are constituted and articulated in societies are an expression of power relations, rather than something evenly produced. Overall, then, the discussion in this chapter aims to take thinking beyond a concern with people's abilities to meet energy needs, to reflect another important dimension of the challenges associated with energy poverty—namely, the processes of evolution, constitution, and enrolment of people in needs for energy services.

In this respect, the analysis can further benefit from looking across to the wellbeing and capabilities literature. This literature highlights how unequal power relations in society mean that people are 'differently able to conceive of, pursue and achieve wellbeing' (Deneulin & McGregor, 2010, p. 513). The analysis here is revealing in that it suggests that people are also differently placed to shape what even becomes important or needed to achieve that wellbeing. That is to say, as new needs for energy take hold in wider society, people are subject to differing requirements to adopt the related practices, with some having little possibility for defection. In the case of welfare, an array of policies simultaneously constitute needs while also reducing abilities to meet them. Within energy poverty research, an important route for future research could be to consider how power relations, marginalisation, and oppression can operate to limit people's ability to *shape needs* as much as it can to meet them once constituted.

The discussion here is further revealing for thinking about how the literature on energy poverty and capabilities opens up important questions for practice-inspired research on energy demand (see also Walker, 2013). Most notably, it foregrounds the importance of inequalities within energy consumption and reveals that processes of recruitment to and defection from practices are coloured by inequality and power relations in important ways. Thinking in these terms opens up a distinctive line of analysis that is engaged more readily with ideas about understanding the place of increasing energy use in creating or further entrenching wider inequalities, bringing to the fore questions about power that are rarely addressed within practice theory-based energy research. Particularly, those regarding who has the power to shape and constitute needs and how abilities to resist, be recruited, or defect from new norms of practice vary across people and across distinct policy areas. This concluding discussion has begun to draw together some of the threads of analysis and argument made through the book, as well as this chapter. The final chapter develops this bringing focus on the key arguments made throughout and offering further reflections on future directions and implications.

References

Bourdieu, P. (1998). *Practical reason: On the theory of action*. Polity Press.

Bouzarovski, S., & Petrova, S. (2015). A global perspective on domestic energy deprivation: Overcoming the energy poverty–fuel poverty binary. *Energy Research & Social Science, 10*, 31–40.

Butler, C., Parkhill, K. A., & Luzecka, P. (2018). Rethinking energy demand governance: Exploring impact beyond "energy" policy. *Energy Research & Social Science, 36*, 70–78.

Citizens Advice. (2020). *Universal Credit: What you need to know as job-hunting rules restart* [online]. https://www.citizensadvice.org.uk/about-us/about-us1/media/press-releases/universal-credit-what-you-need-to-know-as-job-hunting-rules-restart/. Accessed 3 September 2021.

Day, R., Walker, G., & Simcock, N. (2016). Conceptualising energy use and energy poverty using a capabilities framework. *Energy Policy, 93*, 255–264.

Deneulin, S., & McGregor, J. A. (2010). The capability approach and the politics of a social conception of wellbeing. *European Journal of Social Theory, 13*(4), 501–519.

DWP (Department for Work and Pensions) Digital. (2016). *Why are a million digital techies interested in DWP?* DWP Digital Blog [online]. https://dwpdigital.blog.gov.uk/2016/07/29/why-are-a-million-digital-techies-interested-in-dwp/

DWP (Department for Work and Pensions) Digital. (2021). *What digital with purpose means for DWP customers*. DWP Digital Blog [online]. https://dwpdigital.blog.gov.uk/2020/12/08/what-digital-with-purpose-means-for-dwps-customers/

Finn, D. (2018). Why are Britain's jobcentres disappearing? *The Conversation* [online]. https://theconversation.com/why-are-britains-jobcentres-disappearing-91290

Gormally, A. M., O'Neill, K., Hazas, M. D., Bates, O. E. G., & Friday, A. J. (2019). "Doing good science": The impact of invisible energy policies on laboratory energy demand in higher education. *Energy Research & Social Science, 52*, 123–131.

Großmann, K., & Kahlheber, A. (2017). Energy poverty in an intersectional perspective: On multiple deprivation, discriminatory systems, and the effects of policies. In N. Simcock, H. Thomson, S. Petrova, & S. Bouzarovski, *Energy poverty and vulnerability: A global perspective* (pp. 12–32). Routledge.

Mattioli, G. (2017). "Forced car ownership" in the UK and Germany: Socio-spatial patterns and potential economic stress impacts. *Social Inclusion, 5*(4), 147.

Mattioli, G., Lucas, K., & Marsden, G. (2017). Transport poverty and fuel poverty in the UK: From analogy to comparison. *Transport Policy, 59*, 93–10.

Middlemiss, L. (2016). A critical analysis of the new politics of fuel poverty in England. *Critical Social Policy, 37*(3), 425–443.

Middlemiss, L., Ambrosio-Albalá, P., Emmel, N., Gillard, R., Gilbertson, J., Hargreaves, T., Mullen, C., Ryan, T., Snell, C., & Tod, A. (2019). Energy poverty and social relations: A capabilities approach. *Energy Research & Social Science, 55*, 227–235.

Middlemiss, L., & Gillard, R. (2015). Fuel poverty from the bottom-up: Characterising household energy vulnerability through the lived experience of the fuel poor. *Energy Research & Social Science, 6*, 146–154.

Morley, J., Widdicks, K., & Hazas, M. (2018). Digitalisation, energy and data demand: The impact of Internet traffic on overall and peak electricity consumption. *Energy Research & Social Science, 38*, 128–137.

Mould, R., & Baker, K. J. (2017). Documenting fuel poverty from the householders' perspective. *Energy Research & Social Science, 31*, 21–31.

Petrova, S. (2018). Encountering energy precarity: Geographies of fuel poverty among young adults in the UK. *Transactions of the Institute of British Geographers, 43*(1), 17–30.

Robinson, C., & Mattioli, G. (2020). Double energy vulnerability: Spatial intersections of domestic and transport energy poverty in England. *Energy Research & Social Science, 70*, 101699.

Royston, S., Selby, J., & Shove, E. (2018). Invisible energy policies: A new agenda for energy demand reduction. *Energy Policy, 123*, 127–135.

Shove, E. (2002, October). *Rushing around: Coordination, mobility and inequality*. Draft Paper for the Mobile Network Meeting.

Shove, E. (2003). Converging conventions of comfort, cleanliness and convenience. *Journal of Consumer Policy, 26*(4), 395–418.

Simcock, N., Walker, G., & Day, R. (2016). Fuel poverty in the UK: Beyond heating? *People Place and Policy Online, 10*(1), 25–41.

Urry, J. (2002). Mobility and proximity. *Sociology, 36*(2), 255–274.

Walker, G. (2013). Inequality, sustainability, and capability: Locating justice in social practice. In E. Shove & N. Spurling (Eds.), *Sustainable practices: Social theory and climate change* (pp. 181–196). Routledge.

Walker, G., Simcock, N., & Day, R. (2016). Necessary energy uses and a minimum standard of living in the United Kingdom: Energy justice or escalating expectations? *Energy Research & Social Science, 18*, 129–138.

Wright, S. (2016). Conceptualising the active welfare subject: Welfare reform in discourse, policy and lived experience. *Policy & Politics, 44*(2), 235–252.

CHAPTER 7

Conclusions: Reconceptualising Energy Poverty and Practice

Abstract This concluding chapter discusses the key contributions of the book reflecting on the insights produced through the analysis and the wider implications for research, policy, and practice. The significance of the analysis for energy poverty research is discussed, as is the relevance for practice theory-based scholarship on energy. Discussion reflects on the use of the capabilities approach within the analysis and how ideas about invisible energy policy and the constitution of need can be important for understanding energy poverty, but also explores the potential for bringing inequality more firmly into practice theory-based analyses.

Keywords Energy poverty · Practice theory · Invisible energy policy Capabilities

Introduction

This final chapter draws together key contributions of the book. It looks across the different areas of analysis and thinking that have been addressed throughout, drawing conclusions relevant to both energy poverty and practice-inspired energy research. Key tenets of theoretical engagement that have been worked through here concern on the one hand capabilities approaches to energy poverty, and on the other, practice theory-based interventions related to the constitution of need and invisible energy

C. Butler, *Energy Poverty, Practice, and Policy*, Progressive Energy Policy,
https://doi.org/10.1007/978-3-030-99432-7_7

policy. The synthesis and dialogue advanced through the book across these different areas has allowed for distinctive insights to be produced and opened up possibilities for future analysis. In what follows, each section is themed around key areas of contribution. The first discusses core contributions for thinking with energy capabilities approaches. The second moves to examine those arising from engagement with the invisible energy policy agenda in the context of energy poverty. The third is shaped around ways of working with concepts relating to the constitution of need and how they can be revealing for thinking about inequality in practices. Each section builds drawing together the different conceptual advances taken forward within the book and highlighting avenues for future thought and research.

Advancing Energy Capabilities

Though the analytic focus of this book has been on the insights afforded by looking across the energy poverty and practice-based energy research literatures, this has been grounded in a specific approach to energy poverty that adapts capabilities concepts bringing focus on the relations between energy and everyday actions required to achieve a minimally decent quality of life (Day et al., 2016). This first section of the conclusions offers insights relevant to advancing capabilities-based approaches to energy poverty, drawing out key issues and contributions that have arisen through the analysis and wider discussion.

The book takes forward work that has argued for alternative understandings of energy poverty and analyses that place such expanded treatments at their centre (e.g. Bouzarovski, 2018; Bouzarovski & Petrova, 2015; Day et al., 2016; Middlemiss & Gillard, 2015; Middlemiss et al., 2019; Petrova, 2018; Simcock et al., 2016). Key ideas encapsulated in the work of Day et al. (2016) relating to a capability framework for energy poverty have been pivotal in developing a flexible, multifaceted approach to understanding the issues. They suggest a focus on the capabilities that are underpinned by or related to various energy services as an alternative way of thinking about energy poverty and moving beyond heat and efficiency as the central tenets of analysis in this space. This work (and others) offers an expanded way of thinking about energy poverty taking analysis beyond kilowatt hours (kWh) and opening up narrowly defined understandings of energy poverty to encompass insights that can better attune to lived experiences, interconnection, and relationality.

This book has been grounded in these expanded ways of thinking about energy poverty to advance an analysis that is able to consider the multiple ways in which energy is required within daily life to support social participation and wellbeing. However, it also offers contributions to thinking with capabilities-based approaches across different areas of contemporary debate. First, the analysis and discussion here have spoken to debates about the need to distinguish between capabilities that can be understood as essential and those that are not. I argue that rather than seeking ways to specify essential capabilities and related forms of energy use at the outset, these issues might more fruitfully be worked through as part of analyses. By simply being attentive to questions about the essential nature of energy use—rather than attempting to specify uses or thresholds—and examining how it is linked to capabilities through empirical analysis, possibilities are created for important forms of insight. For example, in the analysis in Chapter 5, being alert to notions of essential energy uses and capabilities was revealing for thinking about processes of normalisation associated with experiences of energy deprivation. By interrogating the ways that people normalised experiences of often severe energy deprivation, questions about what essential uses are and how they connect to capabilities were opened up without recourse to either top-down more paternalistic assessments of need or overly narrow conceptions that obscure how processes of oppression shape what people see as acceptable.

One way into analysis, then, that is proposed here is to adopt a biographical methodology (see Butler et al., 2014) as an approach to understanding energy capabilities. Such an approach places the person and their relational context at the centre, rather than a specific form of energy use, sector, or service. It allows for exploration of interconnection between different forms of energy poverty and opens up the analysis to complexity by engaging with lived experiences. I suggest biographical approaches, such as those employed here, can facilitate movement past the traditional spatial boundedness of energy poverty research and offer a route to engaging with capabilities in a grounded way but without the goal being to specify essential needs.

This also connects with debates about the nature of the relations between capabilities and energy services (see Chapter 2; cf. Middlemiss et al., 2019). Middlemiss et al. (2019) assert that while a sequential relationship has been inferred in previous work, their analysis reveals a more bidirectional one wherein energy services can be important to capabilities, but capabilities can also underpin access to such services. Building on this, I have argued for the need to maintain such an openness in analysis of the relations between capabilities and energy services. In this, the analysis has created insights that align with Middlemiss et al.'s contention but also extend this to highlight a cyclical relationship between energy services and capabilities. One example discussed in the book concerned the requirements placed on people to travel and access information and communication technologies. Where these energy service needs could not be met, there were severe implications for people's capabilities relating to securing income, which in turn restricted access to energy services. In this sense, then, in addition to the bidirectionality identified by Middlemiss et al. (2019), such relations can be seen as cyclical.

Beyond this, the analysis in the book has sought to draw out an inherent flexibility within the energy capabilities approach to look beyond the bounded focus on domestic contexts often associated with studies of energy poverty. Though existing work recognises services can be met through means outside of the confines of the home (Bouzarovski & Petrova, 2015; Day et al., 2016), analyses have yet to move to take in mobilities as well. While this is a possibility encapsulated in energy capabilities frameworks, to date little research adopting this approach has worked outside of domestic contexts or moved to look across and between mobilities and domestic forms of energy poverty. By building from an approach to analysis through the book that is grounded in biographical interview data and lived experiences, distinctions between mobilities and domestic energy provision come to be seen as limiting for understanding the ways that different forms of energy deprivation interconnect. The conceptual synthesis here offers an expanded and flexible way of defining energy poverty and combines it with a methodological approach that is focused on lived experience, bringing to light the value of looking at multiple different energy services, including mobilities, their interrelations, and their importance to daily life.

For example, in Chapter 5, data extracts focused around mobilities were revealing for extending thinking about self-rationing and self-disconnection to recognise the ways that people live without some energy services entirely. Within the mobilities space, participants frequently discussed always walking and never using public or private transport because of the associated costs, even where the distances were extremely long and the lack of energy services severely affected capabilities, for example, relating to social respect and securing income. A focus on mobilities has been revealing too for thinking about the processes through which needs are constituted in and through policy with important implications for experiences of energy deprivation and poverty (see Chapter 6). As an area of energy deprivation, it affords insight into how energy poverty can be related to demands for travel that are placed on people through policy, as well as owing to policies that shape options for and costs of travel. Mobilities are perhaps more revealing in this respect precisely because they do not form a focus of fuel poverty policy in the same way as domestic energy use does. The advantages a capabilities perspective affords for looking across different forms of energy use and drawing in mobilities, then, could, I argue, be made far more integral to analyses in this space and afford possibilities for important insights. Overall, applying and utilising a capabilities-based understanding within the book has afforded opportunities to explore possibilities for analysis embedded in this approach and to offer further avenues for its development and expansion within energy poverty research.

Implications for Invisible Energy Policy Analysis

An important starting point for the book was to examine the ways that processes of governance—far beyond energy policy—have implications for the constitution, reproduction, and exacerbation of energy poverty. This arises from engagement with the area of practice-based energy research that has been termed 'invisible energy policy' (Royston et al., 2018). Working with the invisible energy policy approach has brought insights important to understanding the emergence and reproduction of energy poverty, but I want to argue it also offers insight into trajectories for research and analysis taking forward the invisible energy policy agenda.

This research began life as an initial attempt to develop the invisible energy policy idea, focusing on a case example policy area, but what it has revealed has been telling for research more widely. Crucially, by moving beyond the predefined categories, classifications, and distinctions of existing government institutions and focusing on lived experiences and enactments, the intersections and messy realities of policy are revealed (Butler et al., 2018). This approach, then, widens the scope of analysis in important ways for those working on governance both within energy research and beyond. First, it creates space for asking different questions of the processes by which (energy) problems are re/produced and created. Second, it takes analysis beyond definitions and categorisations of the problems as they are currently formulated within existing governance structures. This lends important opportunities for more critical forms of analysis that trouble existing ways of understanding and viewing social challenges.

Developing the analysis from the invisible energy policy agenda has brought focus on the ways that energy poverty and vulnerabilities are actively created and shaped by policy and governance, often emanating from areas other than energy (see also Middlemiss & Gillard, 2015). The focus on welfare policy brings into sharp relief the ways that policy beyond energy policy is deeply implicated in shaping people's abilities to meet their needs and in affecting how those needs manifest. For example, welfare policy (along with other areas of policy) has been shown through this analysis to have implications for where people live and for the inadequacy of housing. Contemporary welfare reforms have also been shown to have had severe impacts on people's abilities to afford energy services and meet basic energy needs, both in the home and in the context of mobilities. The analysis in this book has gone further, however, in considering not only the direct impacts of policy on people, their energy access, and their capabilities, but broader forms of influence that shape experiences of energy poverty in equally important ways.

In this respect, looking across to welfare policy has been revealing for recasting problems and developing lines of analysis within energy poverty in at least two ways. One concerns the fuel poverty-wider poverty divide. Previous research on energy poverty has highlighted the importance of fuel poverty being identified and addressed within policy as a distinctive problem, separate from wider conditions of poverty (Bouzarovski, 2018). The significance (rightly) afforded to such a conceptual and policy distinction has, however, underpinned tendencies to not address the relations between fuel poverty and wider poverty within analyses. By beginning with a different area of policy, the analysis here was poised and pushed to confront such relations allowing for the central significance of energy services in contributing to the cyclical temporalities that characterise poverty (see Macdonald et al., 2020) to be brought to the fore. This underscored, rather than undermined, then, the importance of addressing energy poverty precisely because of its relation to wider poverty.

A second way in which starting from welfare policy has been important for rethinking how problems are typically understood and situated is by bringing a focus on the nature of political subjectivities within different policy areas. In previous work, Middlemiss (2016) has highlighted the different ways that subjects of welfare policy and subjects of fuel poverty policy are positioned within their respective policy areas. Crucially, she highlights how welfare policy is politicised in ways that fuel poverty policy is not. Where fuel poverty subjects are assumed as worthy recipients of policy support and the framings and problematisations tend to be technical, those in welfare policy tend to be cast as unworthy and problems are positioned as arising out of individual deficits, such as in skills, abilities, and willingness.

In Chapter 5, the ways that these different subjectivities play out in the context of lived experiences of both energy poverty and welfare policy are examined. But in these concluding comments I wanted to draw attention to how this type of questioning only arises out of looking beyond the specifics of one policy area. Critical engagement with how political subjects are positioned across different policy areas is important for analyses of governance, but can easily be overlooked with a narrower framing. Crucially, the research and analysis in this book have given focus to the lived experiences of policy allowing for analysis of the ways that these different subjectivities reveal themselves within daily life and are negotiated with consequences for the ways that societal problems take form.

In this case, the ways that the kinds of severe energy deprivation—encapsulated in discussion of 'living without energy'—are accepted and normalised as part of living in poverty have been brought to the fore (see Chapter 5). The subjectivities associated with welfare policy appeared far more dominant than those encapsulated in fuel poverty policy arenas. The analysis reflected on how some of the successes that arise from having fuel poverty as something distinctive and apart from poverty, such as its depoliticisation and policy support, are not necessarily carried through, therefore, to the experiences of energy poverty. It has further highlighted how concerns about eroding the distinction between poverty and fuel poverty in contexts like the UK have perhaps resulted in less attention being given to the ways that energy deprivation is foundational in the creation and cyclical reproduction of wider poverty. Though in Global South contexts these relations might be readily recognised and even core to analysis and policy about energy access (Bouzarovski & Petrova, 2015), they are far more marginal in debates about poverty in the UK and other areas of the Global North.

Starting from the invisible energy policy agenda as a way into the empirical research and analysis has been important, then, in bringing to light insights important to contemporary debates about energy poverty. Concerns relating to the role of wider socio-political structures and processes in institutionalising and normalising energy poverty are brought to the fore, as are issues relating to policy and political definitions of fuel poverty (cf. Middlemiss, 2016; Petrova, 2018). But there are also insights relevant for thinking about research trajectories in the area of invisible energy policy arising from the focus on energy poverty. Engagement with energy poverty research has brought focus on the wider discourses and political subjectivities that shape and are engendered by different policy areas. This focus on the role of different policy areas in shaping energy concerns through broader processes and forms of discursive construction takes analysis beyond specific policies, giving emphasis to alternative routes for analysis of invisible energy policy.

The Constitution of Need in Energy Poverty

The consequences of not being able to meet energy demands or access energy services for wellbeing are of central concern in the energy poverty literature (e.g. Bouzarovski & Petrova, 2015; Middlemiss & Gillard, 2015; Middlemiss et al., 2019; Simcock et al., 2016). In contrast, the nature of increasing needs for energy as part of contemporary life has formed the focus for energy demand research (e.g. Shove, 2003; Shove et al., 2012) but with little consideration of how this relates to inequality and wellbeing. In this context, the ways that ever-increasing needs for energy have implications for energy poverty and wellbeing have been largely overlooked, as have the ways that people are differentially placed to enact new needs within everyday practice or to resist them as they take hold.

By combining these two major areas of conceptual development in energy consumption research, the analysis here has been able to develop the arguments regarding problems with escalating demand to include implications not only for environmental sustainability but also, more directly, for energy poverty and wellbeing. The discussion develops insight into how people have differing degrees of agency in their ability to resist and shape practices, as well as in their capabilities to successfully perform them. This takes thinking beyond a concern with how people with different characteristics are vulnerable to energy poverty, to consider the constitution of requirements for energy use and the ways this is enshrined in policy emanating from different areas of government. This brings in consideration of power and of differential patterns of agency in the enactment and defection from practices, foregrounding questions about governance and intervention in distinctive ways. Crucially, it brings attention to the ways that governance and policy are deeply implicated in constituting needs for energy and, consequently, the conditions for energy poverty. And onto how the move to consider non-energy policy within energy demand sustainability research (see Royston et al., 2018) also has relevance for studies of energy deprivation.

From the analysis I argue that existing inequalities affect people's engagement with practices in important ways. First, existing capabilities afford variable possibilities for subverting, resisting, and reshaping practices constituted, in part, through policies. Consequently, as new forms of practice emerge and become normalised, they produce new forms of exclusion and vulnerability. One of the ways that this is exemplified is in examining the ways that welfare policy has actively constituted new needs for information and communication technologies (ICTs) such that these energy services come to be prioritised above heat and other forms of energy use that are more conventionally the focus of energy poverty. The constitution of needs for ICTs in this context places requirements on people to fulfil these needs or have no way of accessing their income or completing the duties necessary to receive welfare benefits. This highlights the limited agency that exists in some people's abilities to negotiate the enactment of practices (and related energy needs), or to play an active role in shaping the ways that practices are constituted in the first place.

Second, and related to this, is recognition of how different forms of policy act on different people in different ways. For example, different areas of governance are often far more coercive and punitive than others. Within welfare policy, though it is shaped by problematisations of individual behaviour and characterised by forms of governance 'at a distance' (Rose, 1999), it leaves far less latitude for enactments that divert from the dictates of policy. Centrally, this is because any diversions from the requirements of welfare policy are met with punitive sanctions to income and severe hardship. These clear and punishing mechanisms within welfare policy can be contrasted with the less overt mechanisms through which other areas of governance operate. For example, Gormally et al. (2019) show through their research on higher education how the negotiation, and prioritisation, of policies instituted through governance was often, ultimately, aligned with central government priorities (e.g. economic success). Though their research is suggestive of diffuse forms of power operating through people to engender particular kinds of practice, it also suggests a far more negotiated form of engagement with policy in everyday life than is evident in the welfare policy context. The research discussed in this book showed severe limitations on how far policy enactment entailed negotiation and the exercise of power by those subject to it, along with variable capabilities to be able to act in the ways required of them. In Chapter 6, this *enforced* nature of engagement with or 'recruitment' to practices related to energy services was highlighted, as were the ways that people were unable to (successfully) enact these practices in ways that denote full capability.

For energy poverty research, this represents engagement, then, with the other side of the picture, taking in consideration not only of how people's abilities to *meet their needs* are shaped and conditioned by socio-political processes, but of how those *needs are constituted*. Conversely, thinking about the relations between policy and practice in this way brings inequality more firmly into view as important for understanding both invisible energy policy and wider conceptualisations arising from practice theory-based energy analysis. Excepting two conceptual interventions (Shove, 2002; Walker, 2013), there has been only very limited engagement with issues of power and inequality in work addressing energy sustainability from practice perspectives. The analysis in this book brings to the fore the unevenness that exists in people's abilities to both shape the enactment and institutionalisation of practices and emergent 'needs'. It highlights the exclusions and vulnerabilities that are created and exacerbated in the processes through which practices take hold. And it signals something far more draconian and darker that the concept of social norms denotes in the ways that some forms of energy service and related practice are effectively enforced. This, then, brings a more critical politics into the thinking about agency and structure that underpins practice-based approaches to energy sustainability.

Insights for Policy and Wider Responses

For this final section, I turn to reflect on some of the implications for policy and practice related to fuel poverty. The arguments and analysis presented in this book depart markedly from that which forms the current focus for policy in the UK and many other countries. By casting energy deprivation as an issue of capabilities related to energy services, the understanding of energy poverty is taken far beyond that of official definitions that tend to be focused on the domestic context and are principally concerned with heat. Such an understanding of fuel poverty has been highlighted as grounding a specific problematisation (Rose, 1999) that results in an emphasis on building efficiency as *the* approach to addressing the issue (Middlemiss, 2016). Though within policy, account is taken of vulnerabilities (such as being disabled, very young or very old), there is little space for recognition of the many dimensions of the problem that are brought into view by a wider focus on capabilities, quality of life, precarity, and lived experiences (Day et al., 2016; Middlemiss & Gillard, 2015; Petrova, 2018; Simcock et al., 2016). Moreover, there appear few openings for consideration of the ways needs for energy services are (re)produced and constituted through policy in ways that have important effects for energy poverty.

The implications of this research, then, do not sit easily within the existing scope for policy interventions. But this does not preclude opportunities for taking forward some of the implications of this analysis within other spaces of action on fuel poverty. As existing practices by those working in fuel poverty demonstrate, it is possible to both campaign and work outside of the (sometimes) narrow definitions and understandings prescribed by policy and to use (or subvert) existing mechanisms to achieve better outcomes for people living in situations of energy deprivation. Understanding of the importance of energy deprivation for wider cycles of poverty derived from this analysis could form a space for campaigning about the relevance of energy poverty to social inequality challenges more widely, as could the insights relating to the ways that policy can be constitutive of needs for energy (services), while simultaneously limiting people's abilities to meet those needs.

In terms of policy, the frames and ideas that dominate fuel poverty are not currently directed towards such issues of emerging need, but there is no reason to think that this could not be the case. There could be room for a more reflective politics that involves thinking across policy and highlighting intersections that are relevant to the constitution of needs for energy and for abilities to meet those needs. Though this does not in itself represent a clear recommendation, it may be that simply opening up spaces for intersection and processes of constitution to be recognised could be productive of new thinking. This work would be usefully facilitated by a focus on lived experiences as part of both policy and impact analysis.

There might also be greater room for thinking about how other areas of governance intersect to affect the issues within existing policy. The recent Sustainable Warmth Strategy in the UK (BEIS, 2021) moves to consider the relevance of low carbon transitions and sustainability policy for energy poverty, as well as to think about the role of health policy. This suggests openings for insights into the importance of policy areas beyond energy to be at least reflected on within fuel poverty governance. The extent to which energy services and areas of emerging need such as information and communication technologies, or wider energy service priorities, such as transport and mobilities, can be brought more strongly into focus within fuel poverty debates remains an open question. But this research highlights how these areas of need are an important part of the picture in building understanding of energy poverty generally and, particularly of how different energy services intersect and are prioritised.

A clear challenge for policy to be responsive to the more dynamic complex picture of energy poverty created within this book (and the work of others) is the reliance on energy market mechanisms for policy delivery. The primary means through which policy seeks to reduce energy poverty are delivered through private energy companies and this produces a particularly constrained environment in which to address the issues. However, even within this there is much scope for more to be done that would at least be in some way responsive to the severe forms of deprivation delineated within these pages.

With a capabilities-based understanding of energy poverty comes recognition that energy companies could work to address the issues in ways that extend beyond the financial commitment under the Energy Companies Obligation (ECO). For example, in their pricing for poorer households (particularly those on prepayment meters) in their handling of debt and in their communication of options available for managing debt, in their communication more generally so that vulnerable people find it easier to engage, and in the nature of the support they offer for customers living in energy poverty. All this could be improved and expanded, as well as embedded more fundamentally in the culture and approach of energy companies. This, of course, does not take interventions to the places that are suggested by the analysis set out here, but the point is to highlight that much could be done even within the constraints of current approaches and understandings to better support those living without energy.

References

BEIS (Department for Business, Energy and Industrial Strategy). (2021). *Sustainable warmth: Protecting vulnerable households in England* [online]. https://www.gov.uk/government/publications/sustainable-warmth-protecting-v.ulnerable-households-in-england

Bouzarovski, S. (2018). *Energy poverty: (Dis)assembling Europe's infrastructural divide*. Palgrave.

Bouzarovski, S., & Petrova, S. (2015). A global perspective on domestic energy deprivation: Overcoming the energy poverty–fuel poverty binary. *Energy Research & Social Science, 10*, 31–40.

Butler, C., Parkhill, K. A., & Luzecka, P. (2018). Rethinking energy demand governance: Exploring impact beyond "energy" policy. *Energy Research & Social Science, 36*, 70–78.

.Butler, C., Parkhill, K. A., Shirani, F., Henwood, K., & Pidgeon, N. (2014). Exploring the dynamics of energy demand through a biographical lens. *Nature and Culture, 9*(2), 164–182.

Day, R., Walker, G., & Simcock, N. (2016). Conceptualising energy use and energy poverty using a capabilities framework. *Energy Policy, 93*, 255–264.

Gormally, A. M., O'Neill, K., Hazas, M. D., Bates, O. E. G., & Friday, A. J. (2019). "Doing good science": The impact of invisible energy policies on laboratory energy demand in higher education. *Energy Research & Social Science, 52*, 123–131.

MacDonald, R., Shildrick, T., & Furlong, A. (2020). "Cycles of disadvantage" revisited: Young people, families and poverty across generations. *Journal of Youth Studies, 23*(1), 12–27.

Middlemiss, L. (2016). A critical analysis of the new politics of fuel poverty in England. *Critical Social Policy, 37*(3), 425–443.

Middlemiss, L., Ambrosio-Albalá, P., Emmel, N., Gillard, R., Gilbertson, J., Hargreaves, T., Mullen, C., Ryan, T., Snell, C., & Tod, A. (2019). Energy poverty and social relations: A capabilities approach. *Energy Research & Social Science, 55*, 227–235.

Middlemiss, L., & Gillard, R. (2015). Fuel poverty from the bottom-up: Characterising household energy vulnerability through the lived experience of the fuel poor. *Energy Research & Social Science, 6*, 146–154.

Petrova, S. (2018). Encountering energy precarity: Geographies of fuel poverty among young adults in the UK. *Transactions of the Institute of British Geographers, 43*(1), 17–30.

Rose, N. (1999). *Powers of freedom: Reframing political thought*. Cambridge University Press.
Royston, S., Selby, J., & Shove, E. (2018). Invisible energy policies: A new agenda for energy demand reduction. *Energy Policy, 123*, 127–135.
Shove, E. (2002, October) *Rushing around: Coordination, mobility and inequality*. Draft paper for the Mobile Network meeting.
Shove, E. (2003). Converging conventions of comfort, cleanliness and convenience. *Journal of Consumer Policy, 26*(4), 395–418.
Shove, E. Pantzar, M., & Watson, M. (2012). *The dynamics of social practice: Everyday life and how it changes*. Sage.
Simcock, N., Walker, G., & Day, R. (2016). Fuel poverty in the UK: Beyond heating? *People Place and Policy Online, 10*(1), 25–41.
Walker, G. (2013). Inequality, sustainability, and capability: Locating justice in social practice. In E. Shove & N. Spurling (Eds.), *Sustainable practices: Social theory and climate change* (pp. 181–196). Routledge.

FSC
www.fsc.org
MIX
Papier | Fördert
gute Waldnutzung
FSC® C083411

Kolibri 360 GmbH
Ferdinand-Jühlke-Straße 7
99095 Erfurt, Deutschland
produktsicherheit@kolibri360.de